DEPARTMENT OF THE ENVIRONMENT
PLANNING RESEARCH PROGRAMME

CHANGES IN THE QUALITY OF ENVIRONMENTAL STATEMENTS FOR PLANNING PROJECTS

RESEARCH REPORT

1996

REPORT BY:

IMPACT ASSESSMENT UNIT
SCHOOL OF PLANNING
OXFORD BROOKES UNIVERSITY

London: HMSO

© Crown copyright 1996

Applications for reproduction should be made to HMSO'S Copyright Unit,
St Clements House, 2–16 Colegate, Norwich NR3 1BQ

ISBN 0 11 753269 X

Printed on recycled paper: Reprise Matt
Printed in the United Kingdom for HMSO
Dd 302415 C20 5/96

Preface

The research reported in this document was commissioned by the Department of the Environment as a part of its Planning Research programme and was undertaken by staff at the Impacts Assessment Unit, School of Planning, Oxford Brookes University, directed by Professor John Glasson. The project was overseen by a Steering Group comprising representatives of the Department of the Environment, Scottish Office Environment Department, Welsh Office and the Planning Inspectorate. In addition, the Steering Group benefitted from the experience of Karl Fuller, Technical Director of the Institute of Environment Assessment. The research was carried out in the period between January and July 1995.

The views expressed in the report, as well as conclusions and recommendations, are those of the research team and do not necessarily represent the views of members of the Steering Group or the Department of the Environment, the Scottish Office Environment Department, the Welsh Office or the Planning Inspectorate.

Table of contents

	Page No.
Preface and acknowledgements	iii
List of tables and figures	v
Abbreviations	vii
Executive summary	ix

1 Introduction — 1
1.1 Research context — 1
1.2 Quantity and quality of ES activity — 1
1.3 Aims and objectives of the research — 2
1.4 Outline of the research programme and structure of the report — 2

2 Background: review of literature — 5
2.1 Introduction — 5
2.2 Growth in EA activity and ES output — 5
2.3 Nature of ES quality — 9
2.4 Previous studies of ES quality — 12
2.5 Possible determinants of quality — 13

3 Methodology for the review of the matched pairs of Environmental Statements — 17
3.1 Introduction — 17
3.2 Choice of 25 matched pairs of ESs — 17
3.3 Approach to the review of the 25 matched pairs — 20
3.4 Approach to the detailed review of 10 matched pairs — 22

4 Changes in quality: findings from review of 25 matched pairs of ESs — 25
4.1 Introduction — 25
4.2 Specified information/minimum requirements — 25
4.3 IAU criteria — 27
4.4 Other review criteria: Lee and Colley and EU — 30
4.5 Conclusions — 30

5 Perceptions of ES quality in the planning application process: findings from case studies of 10 matched pairs — 33
5.1 Introduction — 33
5.2 Quality of case study ESs — 33
5.3 A semi-quantitative analysis of the ES/EA participants and the planning application process — 35

		Page No.
5.4	Perspectives on changing quality of ESs	37
5.5	Costs and benefits	39
5.6	Themes and issues	40
6	**Determinants of ES quality and changes in quality**	**43**
6.1	Introduction	43
6.2	Determinants of ES quality	43
6.3	Explanation of changes in ES quality	47
7	**Conclusions and recommendations**	**49**
7.1	Conclusions	49
7.2	Recommendations	51

Appendices

1	Bibliography and references	55
2	Initial questionnaire sent to experienced EA practitioners	63
3	Other ES review frameworks	67
4	List of 25 matched pairs: characteristics	73
5	Review framework for the study	75
6	Findings: Lee and Colley, EU criteria	85
7	'Quality for whom?' Case studies: checklist and questionnaire	89

Preface and acknowledgements

The research reported in this document was commissioned by the Department of the Environment, jointly with the Welsh Office and the Scottish Office Environment Department. It was undertaken by the Impact Assessment Unit (IAU), School of Planning, Oxford Brookes University. The research team comprised Prof. John Glasson (project manager), Riki Therivel, Joe Weston, Elizabeth Wilson and Richard Frost.

The research team benefitted from the advice and valuable comments on Progress Reports and Draft Final Report from a Steering Group set up to oversee progress on the research. The members of the Steering Group were John Zetter (Chair - DoE), Jim Burns (DoE), Nigel Barker (DoE), Tom Hardie (Scottish Office Environment Department), Tom Hunter (Welsh Office), Donald Harris (Planning Inspectorate) and Karl Fuller (Institute of Environmental Assessment).

The IAU would also like to thank all the local authority planners, consultants, developers, statutory and non-statutory consultees and voluntary group members who helped in the provision of Environmental Statements (ESs) completed our questionnaires and answered our questions. They provided invaluable help in filling out the picture on the quality of ESs and the Environmental Assessment (EA) process in the United Kingdom, from a range of perspectives. Notwithstanding the advice, help and assistance provided to the authors of the report, any errors of fact and interpretation are theirs. In addition, the views expressed in this report, including the conclusions and recommendations, are those of the authors, and do not necessarily represent the views of the Steering Group, or the Department of the Environment, the Scottish Office Environment Department or the Welsh Office.

Lists of tables and figures

Tables		Page No.
Table 2.1	Example of development plan policies related to ESs	10
Table 2.2	Dimensions in the study of ES quality	12
Table 2.3	Key quality criteria for ESs	12
Table 3.1	25 matched pairs of ESs - project types	19
Table 3.2	10 detailed matched pairs of ESs - project types	22
Table 4.1	Simple 'regulatory requirements' (25 pairs of ESs)	26
Table 4.2	IAU criteria: pre-1991 ESs (25 ESs)	29
Table 4.3	IAU criteria: post-1991 ESs (25 ESs)	29
Table 5.1	Overall assessment of ESs: pre-1991 cases	34
Table 5.2	Overall assessment of ESs: post-1991 cases	34
Table 5.3	Perceptions of meeting minimum regulatory requirements: pre-1991	35
Table 5.4	Perceptions of meeting minimum regulatory requirements: post-1991	35
Table 5.5	Extent to which ES influenced planning decision	37
Table 5.6	Quality of ESs: pre-1991	37
Table 5.7	Quality of ESs: post-1991	37
Table 6.1	ES quality: ESs prepared by consultants or in-house by developer - IAU criteria	45
Table 6.2	ES quality: ESs prepared by the decision-maker or independent applicant - IAU criteria	45
Table 6.3	ES quality v.length of ES (no. of pages)	46
Table 6.4	Changes in the ES experience of developers, consultants and LPAs in the pre-1991 and post-1991 periods	48

Figures
Figure 2.1 | ESs prepared in the UK, July 1988 to September 1994, by type of project, under all regulations | 7
Figure 2.2 | ESs prepared in the UK, July 1988 to September 1994, by regulation under which they were prepared | 7
Figure 2.3 | Distribution of ESs prepared in the UK, July 1988 to September 1994 | 8
Figure 2.4 | Planning & Construction of matched pairs of ESs in study | 8
Figure 3.1 | Location of matched pairs of ESs used in the study | 18
Figure 3.2 | Representativeness of the full population of the 25 matched pairs (EA Planning Regulations; September 1994) | 19
Figure 3.3 | Representativeness of the 10 matched pairs | 22
Figure 4.1 | Coverage of simple 'regulatory requirements': pre-1991 v. post-1991 (25 pairs of ESs) | 27
Figure 4.2 | Marks for IAU criteria: pre-1991 v. post-1991 | 30
Figure 4.3 | Change in ES quality (IAU criteria) within pairs: pre-1991 v. post-1991 | 30
Figure 6.1 | Relationship between 'size' of project and quality of ES | 44
Figure 6.2 | ES quality: consultant and local authority experience (50 ESs) | 46

Abbreviations

BATNEEC	Best Available Technique Not Entailing Excessive Costs
BPEO	Best Practicable Environmental Option
CBI	Confederation of British Industry
CEC	Commission of the European Communities
CPRE	Council for the Protection of Rural England
DG	Directorate General (of CEC)
DoE	Department of the Environment
DoT	Department of Transport
EA	Environmental Assessment
EC	European Communities
EIA	Environmental Impact Assessment
EIS	Environmental Impact Statement
EN	English Nature
ES	Environmental Statement
EU	European Union
ha	hectares
IAU	Impact Assessment Unit
IEA	Institute of Environmental Assessment
IPC	Integrated Pollution Control
LPA	Local planning authority
m^2	metres (square)
MPG	Mineral Planning Guidance (notes)
MW	Megawatts
NRA	National Rivers Authority
NCC	Nature Conservancy Council
PPG	Planning Policy Guidance (notes)
RPG	Regional Planning Guidance (notes)
RSPB	Royal Society for the Protection of Birds
SO	Scottish Office
SOEnD	Scottish Office Environment Department
TCP(AEE)Regs	Town and Country Planning (Assessment of Environmental Effects) Regulations
WO	Welsh Office

Executive summary

1. Introduction

The report sets out the findings of the research commissioned by the Department of the Environment (DoE) and undertaken by the Impact Assessment Unit (IAU) of the School of Planning at Oxford Brookes University. The principal aim of the research is to establish clearly what changes, if any, there have been in the quality of Environmental Statements (ESs) since the earlier DoE commissioned study which covered the period July 1988 to December 1989 (DoE, 1991a). The focus of the research is limited to ESs produced for projects which require applications for planning permission under the Town and Country Planning (Assessment of Environmental Effects) Regulations 1988 and the Environmental Assessment (Scotland) Regulations 1988. This approach covers approximately 75% of the ESs produced in the UK to date.

2. Background

A review of over 150 items of published and semi-published literature, paralleled by a small and targeted survey, by postal questionnaire, of 16 of the most experienced EA practitioners in consultancies, industry and local government in the UK, provided a guide to the growth of EA activity and ES output, and to the nature and determinants of ES and EA quality. IAU estimates indicate that approximately 2300 ESs had been prepared in the UK between July 1988 and September 1994. Approximately 10% were for Schedule 1 projects and 90% for Schedule 2 projects. The largest number by project type was for waste disposal, accounting for 22% of all ESs. Roads, industrial and urban projects, extraction and energy projects also accounted for large numbers of ESs.

Key criteria for ES quality were identified as compliance with regulations, comprehensive of information, adequacy of methodology, clarity and organisation of information, effective communication and accessibility to relevant audiences, and transparency, objectivity and impartiality. Relevant determinants of ES/EA quality include: the type and size of the project, the experience of the developer, consultant and local authority, and EA related guidance and factors related to the planning process.

The review of the literature, informed by the survey of key practitioners, also suggested that a study of ES quality should include several dimensions. It should consider both simple 'minimum regulatory requirements' and more comprehensive 'best practice' criteria; it should recognise the importance of setting any discussion of ES quality within the wider context of EA quality and of the EA and planning application process; and it should consider 'quality for whom', examining the various perspectives on ES quality. Such considerations influenced the methodology used in the study.

3. Methodology

Changes in ES quality, over the periods 1988-1990 to 1992-1994, were assessed at two levels. The 'macro study' included the classification of the full populations of projects for which ESs had been submitted, under the EA Planning Regulations, for the two study periods: pre-1991 and post-1991. From these two population sets, 25 pairs of projects (50 ESs in total), matched on several criteria, were chosen for detailed ES review. The review included the use of simple 'regulatory requirements' criteria and, in contrast, a very comprehensive review framework developed by the IAU team. In addition, the 25 matched pairs were reviewed using the Lee and Colley review framework (1991) and a European Union (EU) checklist (CEC, 1993).

The second stage of the review methodology constituted a 'micro study' for a sub-set of 10 matched pairs of ESs, covering the two study periods. It focused on case files and involved telephone and direct interviews with the participants involved in the production and processing of the ESs, using a structured questionnaire, to ascertain various perspectives on both ES and EA quality. This second stage looked more at the overall EA process. It also provided a vehicle to undertake parallel research on identifying the costs (and benefits) associated with the production of ESs, and developing a methodology for assessing such costs.

4. Changes in ES quality: research findings

The 'macro study' showed that there had been an improvement in the quality of ESs, over the quite short period of time reviewed in the research, whichever of the review criteria were used. For the simple 'regulatory requirements', 44% of the post-1991 ESs fulfilled all the nine criteria used, compared with 36% of the pre-1991 ESs. A more detailed analysis indicated that 92% of the post-1991 ESs fulfilled six or more of the criteria, compared with 64% of the pre-1991 ESs. Using the more comprehensive range of criteria established by the IAU, the overall quality of ESs rose from just unsatisfactory pre-1991, to just satisfactory post-1991. The percentage of satisfactory ESs increased from 36% to 60%. Whilst it is pleasing to note the improvement in quality, there must still be concern that many of the post-1991 ESs, from between one third and one half according to the criteria used, were still unsatisfactory, and in several cases poor.

Quality is also partly in the 'eye of the beholder', and the emphasis in the 'micro-study' was on 'quality for whom' including, in particular, the perspectives of the local planning officer, the statutory and non-statutory consultee, and the developer and consultant. From the evidence of those interviewed, and from background case files, there was a perception from several perspectives that there had been an improvement in ES quality over the two study periods. There was also a clear view from all parties that pre-application activities, with early consultation, negotiation and participation in the scoping of the ES, were important to the quality of the ES, and to the time taken in considering the application.

The 'micro-study' also highlighted the reliance of LPAs on consultees to assess various elements of the ES post-submission. The importance of reputable consultants in the production of a good quality ES was another finding. There was little quantitative evidence of the costs or time taken in processing planning applications accompanied by an ES; however there is some evidence that good quality, comprehensive, statements can bring resource savings.

5. Determinants of the changes in quality

High quality is associated with more experienced developers, consultants, consultees and LPAs. Experience has increased for all participants in the process. Higher quality has been associated in particular with the 'good' environmental consultant; although there is concern about the dangers of complacency, and overstandardisation, perhaps as a result of price competition. There is also concern about the opportunistic new entrant to the field with little to offer other than a low price. Overall, the expectations of ESs from all parties, including pressure groups, are rising over time and this is a force for improved quality.

Improvements in guidance and training may also explain some of the improvements in quality. There are also more ESs in the public domain to provide evidence of good practice. Whilst not particularly subject to policy influence, a change towards projects normally associated with good ESs and away from those normally associated with poor ESs, plus the increasing familiarity with more recent project types, may also lead to an improvement in the quality of the population of ESs.

6. Recommendations

The study has revealed some improvements in the quality of ESs from those produced pre-1991 to those produced post-1991. It has also revealed that a substantial proportion of ESs are not of satisfactory quality, and that within the content of the ES, and within the wider EA procedures, several problem areas can be identified. The report concludes with a number of recommendations in response to particular issues raised in the study. These include recommendations to:

- improve the extent and nature of pre-submission consultation and scoping activities;
- encourage the dissemination of good practice guidance by central and local government, with a focus on areas of particular weakness in ESs, including the consideration of alternatives and the nature of the non-technical summary;
- facilitate the recognition and use of good quality practitioners;
- encourage the use of independent review of ESs prior to submission, and the use of review frameworks by LPAs; and
- improve the quality of ESs in terms of effective communication and accessibility to relevant audiences, through an improved and free-standing non-technical summary, and the provision of ESs at reasonable price.

Recommendations are also made in relation to associated issues. These relate in particular to improving the availability and specification of the ES document.

The recommendations in this report are set in the context of EC consideration of amendments to the EA Directive, which, if and when implemented, will undoubtedly require some changes in the UK's EA and planning system.

Ideally the EA process should be an iterative and dynamic process which provides the opportunity of improving projects and safeguarding the environment at every stage. The UK planning system has so far accommodated the introduction of this process and the changes it has brought. The evidence from the study is that it has been a force for good; as ES quality and EA practice improve and develop so will the benefits to the planning process and environment increase.

CHAPTER ONE

Introduction

1.1 Research context

The European Community (EC) Directive 85/337/EEC came into force in July 1988. For certain types of project which are likely to have significant environmental effects, the Directive requires an Environmental Assessment (EA)[1] to be carried out before (development) consent is given. In such cases, the developer is required to produce a report describing the likely effect the project may have on a number of environmental factors. This report is referred to as an Environmental Statement (ES)[1].

In the UK, the EC Directive has been implemented through a series of Regulations. For projects requiring development consent from a local planning authority (LPA) in England and Wales, the Directive has been given legal effect through the Town and Country Planning (Assessment of Environmental Effects) Regulations (DoE, 1988b); in Scotland it has been given effect by the Environmental Assessment (Scotland) Regulations. The EA Planning Regulations apply to two separate lists of projects, based on Annexes I and II to the EC Directive:

- Schedule 1 projects, for which EA is required in every case; and

- Schedule 2 projects, for which EA is required only if the particular development is likely to give rise to significant environmental effects.

Advice to LPAs on procedures and on the implementation of the EA Planning Regulations is set out in Official Circulars published by the DoE, Welsh Office and the Scottish Office (DoE, 1988a). Initial guidance for developers and others was published in a booklet *Environmental Assessment: a Guide to the Procedures* (DoE, 1989).

1.2 Quantity and quality of EA/ES activity

Environmental Assessment is an approach in good currency. Since the implementation of the EC Directive in the UK in 1988, over 2000 Environmental Statements have been produced, and there has been a significant increase in the annual quantity of ESs produced since 1988 (Frost and Frankish, 1995). Approximately 75% of the ESs have been produced under the EA Planning Regulations. It might reasonably be anticipated that this level of activity would also lead to improvements in the quality of ESs also, as the various parties involved - developers, consultants, LPAs, statutory and non-statutory consultees, other interest groups and Central Government - improve their knowledge, understanding and skills of EA, and move up the learning curve.

An early study of ES quality was included in a report produced by the University of Manchester for the Department of the Environment, jointly with the Welsh Office and Scottish Office. This report *Monitoring Environmental Assessment and Planning* (DoE, 1991a), covered the first 18 months of the operation of the formal EA system. One of the findings of the report was that the majority of ESs reviewed were adjudged to be of 'inadequate' quality. The report made recommendations about how to address this and there has been subsequent action, including the recent publication of *Good Practice Guides* on *The Evaluation of Environmental Information for Planning Projects* (DoE, 1994d/e) and on *Preparing Environmental Statements for Planning Projects* (DoE, 1994g), though these and other procedural measures have yet to take effect fully.

A further review of recent practice has been undertaken by the European Commission, to assess the implementation of EC Directive 85/337 (CEC, 1993a). This review welcomes the introduction of common legislation across Member States, the provision of information on projects, and the general spread of good practice. But there is concern about the opaque processes involved, the limited access of the public and the lack of continuity in the process. The EC notes that 'adequate control of the ES and of the EA process as a whole is not always present', and that, in the UK 'measures should be taken to improve the quality and objectivity of ESs' (CEC, 1993a).

1.3 Aims and objectives of the research

The principal aim of this research is to establish clearly **what changes**, if any, there have been in the quality of ESs since the earlier study by the University of Manchester (DoE, 1991a), and to explain **why such changes**, if any, have taken place. A secondary aim is to establish the feasibility of **identifying additional costs** to relevant parties attributable to the production of the ES, and to develop a methodology for assessing such costs. The report on this secondary aim is separate from this report, but the aim is noted here as it has some bearing on the research methodology used.

The focus of the research is limited to ESs produced for projects which require applications for planning permission under the Town and Country Planning (Assessment of Environmental Effects) Regulations 1988 (with subsequent amendments) and the Environmental Assessment (Scotland) Regulations 1988. Whilst this approach covers a large proportion of ESs produced to date, and facilitates comparison with the earlier study, it does remove some potentially interesting comparisons with the output from other procedures, some of which have the reputation for being innovative (e.g. ESs for projects under the Electricity and Pipeline Works (Assessment of Environmental Effects) Regulations of 1989).

1.4 Outline of the research programme and structure of the report

The report is structured to reflect the main elements of the research programme.

Chapter 2 presents the findings from a review of the published literature in relevant texts and journals on the quality of ESs in England, Scotland and Wales, and possible reasons for changes in quality. The review was paralleled by a small survey, of some of the most experienced EA practitioners in consultancies, industry and local government in the UK. The aim was to cover also relevant semi-published or in-house literature on the changing quality of ESs, and to identify a variety of perspectives on the nature of EA and ES quality, and on determinants of that quality. This initial review stage also included the specification of an operational definition of 'quality' to be used in the study; and criteria and review frameworks to be used to assess ES performance.

Chapter 3 outlines the two stage review methodology used in the research. This included the classification of the full populations of projects for which ESs had been submitted, under the EA Planning Regulations, for two periods: pre-1991 and post-1991. From these two population sets, 25 pairs of projects (50 ESs in total), matched on several criteria, were chosen for detailed ES review. The review included two main frameworks: a simple regulatory requirements approach and, in contrast, a very comprehensive review framework developed by the project team. In addition, the 25 matched pairs were reviewed using the Lee and Colley review framework (1991) and a European Union (EU) checklist (CEC, 1993). The second stage of the review methodology focused on case files for 10 of the matched pairs, and involved telephone and direct interviews with the participants involved in the production and processing of the ESs, using a structured questionnaire, to ascertain various perspectives on ES and EA quality. This second stage looked more at the overall EA process.

Chapter 4 brings together the findings on changes in ES quality from the review of the 25 matched pairs. The ESs' coverage of the simple regulatory requirements are discussed first, followed by coverage of the criteria developed by the Impacts Assessment Unit (IAU) project team. The findings are presented in aggregate, and are also disaggregated to assess changes in quality for particular elements of the ES.

Features of good and poor ESs are highlighted. Findings are also provided using the Lee and Colley framework, to provide continuity and comparison with the earlier University of Manchester study, and using the recently devised EU checklist.

Chapter 5 discusses the findings from the detailed case studies of 10 matched pairs of ESs. These include the role of the ES and EA in the planning application process from various perspectives, and discussion of the quality of the ES and the costs and benefits of its production for the LPA, developer, consultant, statutory and non-statutory consultee and others involved in the process.

Chapter 6 discusses the explanatory value of various factors in relation to the findings from the review of the 25 matched pairs, and from the detailed case studies, with regard to level of ES quality and changes in ES quality from pre-1991 to post-1991. This includes factors related to: the projects, the participants in the EA process, and other elements of the EA process, including guidance.

Chapter 7 brings together the findings from the literature review and expert survey, and from the 'macro study' of the 25 matched pairs and the 'micro-study' of the case files and interviews for the 10 matched pairs, to draw conclusions on changes in quality, determinants of change, and the perspectives of the various participants in the process. Recommendations are proposed in relation to a number of key issues which have been highlighted in the studies.

---

1. The UK terminology of EA and ES will be used in this report, for consistency with UK regulations, in contrast to the more internationally recognised terms of Environmental Impact Assessment (EIA) and Environmental Impact Statement (EIS)

CHAPTER TWO

Background: review of literature

2.1 Introduction

The literature review provides a background and basis for subsequent stages of the study. It draws on published and semi-published literature on ES quality; a bibliography at Appendix 1 provides reference to over 150 relevant publications on ES/EA quality. The review was paralleled by a small and targeted survey, by postal questionnaire, of 16 of the most experienced EA practitioners in consultancies, industry and local government in the UK. The aim was to cover also relevant semi-published or in-house literature on the changing quality of ESs, and to identify a variety of perspectives on the nature of EA and ES quality, and on determinants of that quality. Valuable responses were received from 13 of the practitioners. Details of this initial questionnaire and the respondents are included in Appendix 2. The chapter begins with a brief update on the growth of EA activity and ES output in the UK, before discussing the nature of ES quality, previous studies of ES quality, and possible determinants of quality.

2.2 Growth in EA activity and ES output

2.2.1 Key participants in EA activity

Key participants in the EA and associated planning and development process include: the developer; the competent authority (for planning projects, normally the LPA); central government; statutory and non-statutory consultees; interest and specialist groups, and the general public, and various facilitators, including planning and environmental consultants, advisers and advocates. With the growth in EA activity post-1988, the number of participants within these various groups has increased. For example, by 1994 it is estimated that over 80% of competent authorities in the UK had received at least one ES (Frost and Frankish, 1995). Partly reflecting the growth in EA activity, there has also been a major increase in the number of environmental consultancies, from approximately 200 in the late 1980s to over 400 in 1995 (ENDS, 1995).

2.2.2 Sources of ES output

There is no one central collection of all ESs produced in the UK to date. When an English LPA receives an ES, it is required to send a copy to the regional office of the Department of the Environment (DoE), which then forwards it to the DoE Library in London when the application is completed. The Library is open to the public on an appointment basis. However this process is a long one, and as a result the collection of ESs held in the DoE Library is partial. In the absence of a comprehensive collection, several other organisations have sought to establish collections and data bases. These include: the Impact Assessment Unit (IAU) at Oxford Brookes University, with a 1995 collection of approximately 600 ESs, and the EIA Centre at

Manchester University, with a 1995 collection of approximately 450 ESs. Both collections are open to the public on an appointment basis. In addition, the IAU produces an annual *Directory of ESs* giving summary details of most ESs produced to date (Frost and Frankish, 1995).

The Institute of Environmental Assessment (IEA) based in Lincolnshire has published a *Digest of Environmental Statements* (IEA, 1993), which provides comprehensive summaries of over 1000 ESs. Updates are supplied on an instalment basis. The IEA also has a collection of about 450 ESs. Until recently the *Journal of Planning and Environmental Law* listed ESs (produced under the Planning EA Regulations) received by the DoE, Scottish Office and Welsh Office on a quarterly basis, giving the project type, LPA and planning decision. It has probably been the most consistently up-to-date list available, but it is not comprehensive, and in future it will only receive copies of Secretary of State 'directions'. Finally, various other organisations such as the Royal Society for the Protection of Birds (RSPB), English Nature (EN), the National Rivers Authority (NRA) and various environmental consultancies have collections of ESs. These are normally not open to the public.

2.2.3 Numbers and types of ESs

From its annual surveys of competent authorities, the IAU has estimated that approximately 2300 ESs had been prepared in the UK between July 1988 and September 1994. Whilst this figure is neither totally comprehensive nor accurate, it gives an indication of the level of EA activity since the implementation of the EC Directive. On average, county and regional LPAs have received considerably more ESs (average 12) than other (district, unitary etc) LPAs (average 3).

Figures 2.1 and 2.2 show the types of project for which ESs have been prepared and the relevant regulations. Figure 2.1 includes both Schedule 1 (approximately 10%) and Schedule 2 (approximately 90%) projects together. The largest project type is that of waste disposal, made up largely of landfill/raise projects, wastewater/sewage treatment schemes, and incinerators. Of the other categories, extraction schemes are largely sand and gravel projects, with the remainder primarily opencast coal. Windfarms constitute the largest group of energy projects. Figure 2.2 shows that approximately 75% of the ESs submitted to date have been under the Town and Country Planning (Assessment of Environmental Effects) and Environmental Assessment (Scotland) Regulations. It is this major group which is the focus of this study. The distribution of project types within this large sub-set is detailed in Figure 3.2(a).

2.2.4 Location of projects

Of the 2300 ESs produced by September 1994, approximately 79% have been submitted in England, 11% in Scotland, 8% in Wales and 2% in N. Ireland. Figure 2.3 shows the distribution of ESs submitted by county and Scottish region to September 1994. The high number for Kent reflects Channel Tunnel and associated activity. The major conurbations also have high numbers. The types of development vary considerably between regions reflecting differences in the local economic base. For example, ESs for opencast coal schemes are particularly prevalent in Derbyshire and the northern English counties, windfarms in Wales and Cornwall, and sewage treatment schemes in the Scottish Highlands, the West Country and Kent.

2.2.5 Planning and construction status of projects

Figure 2.4 shows how many projects of the total in each sector have actually started on the ground. The figure can only be indicative because in only about 50% of the cases has the decision status been determined (either because the decision is still pending or because the result is unknown). Combining the proportion of the block where work has begun with the approved (but not started) segment gives the total number approved. In comparison to other sectors it appears that industrial and urban projects have the highest refusal rates, whereas roads have the lowest.

Figure 2.1: ESs prepared in the UK, July 1988 to September 1994, by type of project, under all regulations.

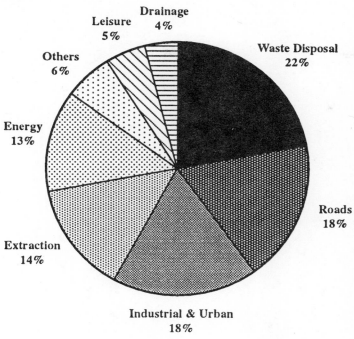

Roads	Special Roads	1.7		Energy	Major Power Stations	1.2
	Local Highway	2.10d			Minor Power Stations	2.3a
					Windfarms	2.3k
WasteDisposal	Landfill/Landraise	2.11c			Pipelines	2.10h
	Sewage/WasteWaterTreatment	2.11d			Transmission Lines	2.3b
	Special Waste Incinerator	1.9				
	Landfill (special waste)	1.9(2)		Leisure	Hotels & Holiday Villages	2.11a
	Recycling Schemes	N/A			Marinas	2.10j
					Golf Courses	N/A
Extraction	Sand & Gravel	2.2c				
	Opencast Coal	2.2d		Others	Intensive Livestock	2.1b,c
					Food Industry	2.7
Industrial & Urban	Industrial Estates	2.10a			Reservoirs	2.10f
	Urban Development	2.10b			Afforestation Schemes	N/A
	Metal Working	2.4			Motorway Service Areas	2.10k
	Chemical Industry	2.6				
	Timber & Paper Industry	2.8		Drainage	Flood Relief	2.10e
					Coastal Protection	2.10l

Figure 2.2: ESs prepared in the UK, July 1988 to September 1994, by regulation under which they were prepared.

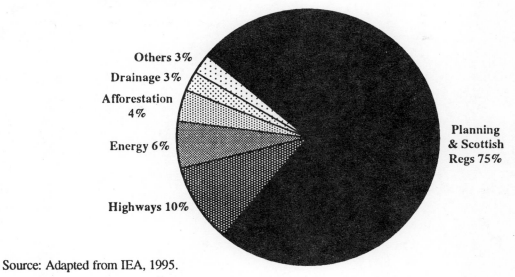

Source: Adapted from IEA, 1995.

Figure 2.3: Distribution of ESs prepared in the UK, July 1988 to September 1994.

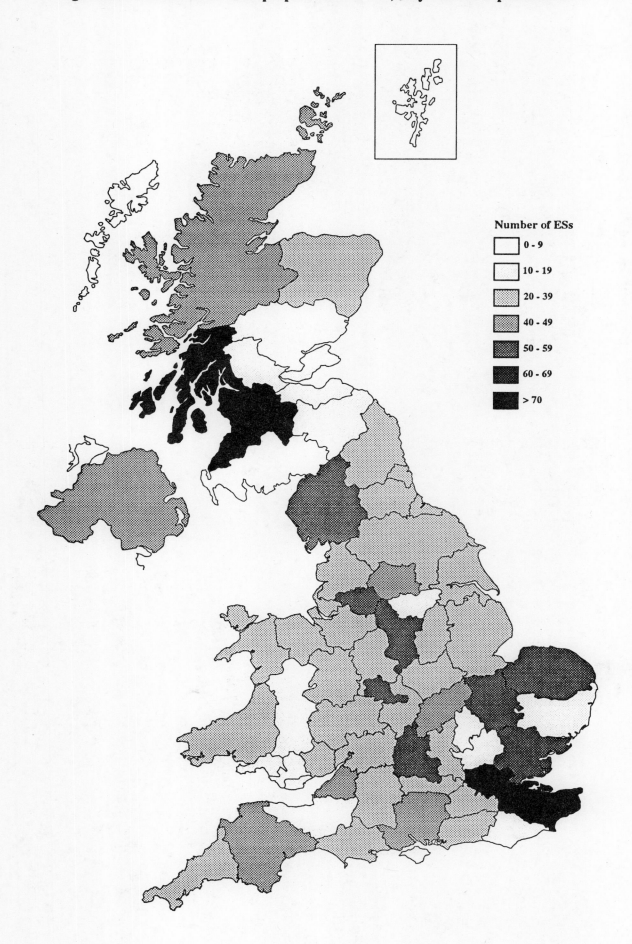

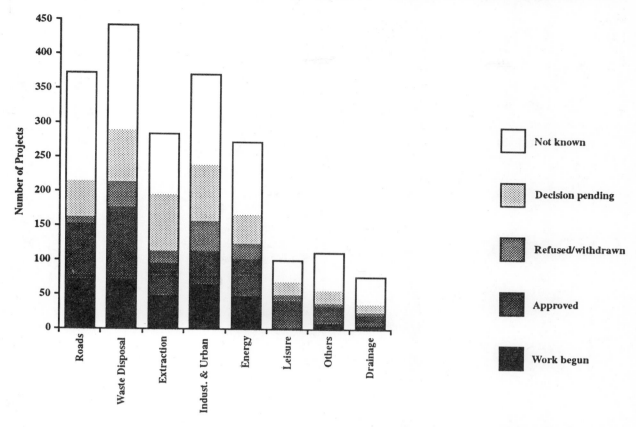

Figure 2.4: Planning and Construction status of ES projects in the UK, July 1988 to September 1994.

2.3 The nature of ES quality

2.3.1 Context: legislation and policy guidance

One approach to quality is to assess whether an ES complies with regulatory requirements. For planning projects, these are the 'specified information' requirements of Schedule 3 of the Town and Country Planning (Assessment of Environment Effects) and the Environmental Assessment (Scotland) Regulations, 1988. These are sometimes referred to as the 'minimum regulatory requirements' of an ES (Glasson et al, 1994). There is evidence of widespread use of this approach by LPAs, as analysed by Land Use Consultants (DoE, 1994d), and in legal comments on and interpretations of the application of the Regulations (see for example, Carnwath, 1991; Scottish Office, 1992; Alder, 1993). The research report, *Good Practice on the Evaluation of Environmental Information for Planning Projects* (DoE, 1994d) amplifies the five criteria in Para. 2 to Schedule 3 of the UK TCP(AEE) 1988 Regulations to provide a checklist of eight criteria to determine the basic adequacy of an ES. The 'regulatory require-ments' specification used in this study (see Para. 3.3 and Appendix 5) is similarly based, and uses a nine criteria checklist.

However, the regulatory requirements are not the only influence on the system. Various policies are likely to influence the content and quality of ESs. Chief amongst these will be the policies of the relevant development plan and the policies of national government as set out in Planning Policy Guidance Notes (PPGs), Regional Planning Guidance Notes (RPGs), Mineral Policy Guidance Notes (MPGs) and DoE/WO/SO Circulars. There has been considerable policy change in recent years, much of which has raised the profile of, and quality expectations for, ESs.

Examples of this post-1988 policy change include the introduction of concepts such as Integrated Pollution Control (IPC), Best Available Technique Not Entailing Excessive Costs (BATNEEC) and Best Practicable Environmental Option (BPEO), associated with the 1990 Environmental Protection Act. PPGs 1, 4, 7, 13 and 20-24 all have sections on EA, as do many of the MPGs. Another important change since the implementation of the EA Regulations is the primacy of the development plan introduced by the 1991 Planning and Compensation Act. 'The increasing emphasis

Table 2.1: Example of development plan policies related to ESs.

> "Notwithstanding the provisions of the Town and Country Planning (Assessment of Environmental Effects) Regulations 1988, the City Council will require the preparation of Environmental Statements for any development which it considers could have significant effects on the environment by virtue of its nature, size and location.
>
> The City Council will require Environmental Statements to be prepared for the following developments:
> - Tipner Redevelopment (Policy GS6)
> - "Energy from Waste" Plant (Policy E58)
> - Ferry Port Extension (Policy ED27)
> - Eastney Waste Water Treatment Works (Policy LC46)"
>
> (Portsmouth City Local Plan, Deposit Version, May 1992).

on the role of development plans in guiding development control decisions requires that ESs should give careful consideration to all relevant policies and plans' (Nelson, 1994). Indeed, some plan policies now refer directly to what issues are expected to be covered in particular types of ESs; Table 2.1 provides an example.

There is also increasing recognition of, and commitment to, sustainable development. This has evolved through various reports, from *This Common Inheritance* (HMG, 1990) to *Sustainable Development: the UK Strategy* (HMG, 1994). These reports recognise the role of EA in contributing to sustainable development, and raise the EA profile amongst all the key user groups. Other policies within PPGs and Circulars, for example PPG1 *General Policy and Principles* (DoE, 1992b), also reflect government commitment to sustainable development, and may similarly raise the EA profile.

2.3.2 ES quality v. EA quality
The UK EA regulations refer specifically to an 'Environmental Statement'. In the EU Directive there is no mention of an ES, only 'Environmental Information' which UK guid-ance takes to include: the ES, any consultations based on it and any further information submitted (DoE, 1989a). Some commentators (eg. CPRE, 1991) have argued that over-emphasis on the ES, and in particular on ES quality, has diverted attention away from the effectiveness of the overall EA process. Street (1993) believes that the ES represents more of a starting point than an end point for negotiations over the environmental design of the project between developers, LPAs and other consultees. A recent DoE research report (1994d) on the evaluation of environmental information has distilled elements of this debate that relate to the issue of quality review in EA.

Two schools of thought exist about the standards which should be required of the initial ES. Some would argue that developers should be encouraged to submit ESs of the highest standard from the outset. This reduces the need for costly interaction between developer and LPA (Ferrary, 1994), provides a better basis for public participation (Sheate, 1994), places the onus appropriately with the developer and increases the chance of a good EA overall. Others would argue that the advice of statutory consultees, comments of the general public, and the expertise of the competent authority can substantially overcome the problems of a poor ES (Braun, 1993). There is also support for this view from planning inspectors at appeal and judicial review cases. This study covers both schools of thought, with a focus primarily on the quality of the ES in the 'macro-study', but with a much wider perspective on quality in the EA process in the 'micro-studies'.

2.3.3 Quality for whom?
The initial review, in this study, of expert opinion on key quality criteria highlighted in particular the objectivity of the ES, its usefulness to the decision maker and its accessibility to various audiences. Petts and Eduljee (1994) argue that ES quality will be determined by the quality of the EA process and by the quality of communication, which 'has to be tested against the needs of the users of the ES'. These needs are 'similar in terms of requiring an objective,

clear and structured explanation of the proposal and assessment of impacts', but also differ in that:

> 'for example for those with specialist responsibilities, interests and training, there is a requirement for detailed technical and scientific information and analysis. For those who lack specialist training, there is a need for non-technical information which is clearly presented.'

Whilst perspectives on quality may vary between groups, they may also vary within groups. Some **developers** may see the design and efficient decision making benefits of a good ES; others may reflect more the views of the CBI Planning Task Force in its publication *Shaping the Nation* (CBI, 1993), which highlights the problems of cost, timescale and conflict. For **Local Planning Authorities**, recent surveys suggest that ES quality is defined pragmatically in terms of its utility in determining the planning application. Cowan (1994) for example suggests that planners use the ES to target a 'shopping list of issues'; Hammersley (1994) argues that LPAs are more likely to take a general overview. Wood (1994b) suggests that the ES has a variable utility in Planning Committees; 20% of the planning officer's report to Committee failed to mention the relevant ES, but in 33% of cases the ES was 'very important' in influencing the Committee decision.

Statutory and non-statutory consultees have a key role in advising LPAs on ES content and quality. Certain national environmental pressure groups such as the Council for the Protection of Rural England (CPRE) have been critical of the quality of ESs (Sheate, 1994), particularly in the areas of scoping of impacts, assessing those impacts and involving the public. The views of the **general public** of the quality of ES are hard to determine. The presentation of information is particularly important for the non-specialist. What first impressions does the ES give? Fuller (1994) clearly advocates such considerations as quality criteria:

> 'If the document appears to be a daunting prospect, then this is an important finding. One of the functions of the ES is to convey information to the non-specialists. A document which immediately puts the reader off is failing to adequately communicate the information.'

Concern amongst **professional consultancies** about standards for EA practice and ES quality led to the establishment of, for example, the Institute of Environmental Assessment (IEA), and the development of guidelines for particular topics in the preparation of ESs, such as those for Landscape and Visual Impact Assessment (IEA, 1995). Those involved in preparing ESs may also apply their own quality control mechanisms such as BS5750 and ISO9000.

2.3.4 Quality criteria

Ortolano (1993) identified five major approaches to EA, which in turn can be interpreted in terms of ES/EA quality:

(a) compliance with applicable regulations;
(b) completeness with respect to the agreed scope of the EA work to be undertaken;
(c) adequacy of the methods used with regard to guidelines, peer review, judicial review etc.'

(d) influence on the weight given to environmental factors in decision making; and
(e) influence on the decision of whether to approve, reject or modify the project.

Other possible criteria include (f) its cost-effectiveness, and (g) its contributions to sustainable development (Sadler, 1994). Of these criteria, only the first three (a-c) apply to the ES rather than to the entire EA process.

A number of ES review criteria have been used to date in the UK, ranging from simple compliance with relevant regulations, to more complex approaches. Three examples are briefly noted here: 'regulatory requirements', Lee and Colley (1990) review package, and a review checklist prepared for the European Commission's Directorate General XI, fuller details of which are included in Appendix 3. The regulatory requirements approach was noted in section 2.3.1. It involves compliance with nine criteria, amplified from Para.2 to Schedule 3 of the Planning AEE Regulations. The review criteria published by Lee and Colley are used by some participants in the EA process. They review ES quality across four areas in a tiered system of criteria. They have also been adapted by the IEA for their own work on ES quality. The EU criteria are more recent (CEC, 1993b). Prepared by Environmental Resource Management, they review the completeness and suitability of environmental information from a technical and decision making viewpoint. The EU review checklist organises criteria in eight review areas; it has not been widely used to date in the UK, because it has only recently been introduced.

To conclude on the nature of ES quality, the

Table 2.2: Dimensions in the study of ES quality.

- consider both 'minimum requirements' and 'best practice' approaches to ES quality;
- recognise the importance of setting any discussion of ES quality within the wider context of EA quality and the EA and planning application process; and
- consider quality for whom, with the various perspectives on ES quality.

Table 2.3: Key quality criteria for ESs.

- compliance with regulations;
- comprehensiveness of information;
- adequacy of methodology;
- clarity and organisation of information;
- effective communication and accessibility to relevant audiences; and
- transparency, objectivity and impartiality.

literature review, informed by the survey of key practitioners, suggests that (i) a study of ES quality should include several dimensions (Table 2.2) and (ii) an operational definition of ES quality should include several key criteria (Table 2.3).

2.4 Previous studies of ES/EA quality

2.4.1 Methods and issues of analysis

There have been a number of studies of the changing quality of ESs/EA in the UK. Several, emanating from the EIA Centre, University of Manchester, have a degree of consistency of approach, using the Lee and Colley review criteria (Appendix 3). Other studies also provide useful overviews, although approaches to the choice of sample, time period and the varying perspectives of the assessors, complicate the comparison of findings. A distinction is made here between aggregated and disaggregated approaches to the assessment of quality.

2.4.2 Aggregated approach

Studies by Wood and Jones (1991) of 24 ESs produced under the EA Planning Regulations between July 1988 and December 1989 suggested that 37% of ESs were of satisfactory quality and 63% were not. A later study by Lee, Walsh and Reeder (1994), of a different set of 47 ESs again produced under the EA Planning Regulations, showed a shift from 17:83 satisfactory : unsatisfactory (1988-89) to 47:53 satisfactory : unsatisfactory (1990-91). Studies of ESs under all regulations (Lee and Brown, 1992), suggest a similar improvement in quality, although the time intervals used are very short. The Lee and Brown study shows also that ESs prepared under the Planning Regulations were only of intermediate quality when compared with those prepared under the Electricity and Pipeline Works Regulations (which were of above average quality) and the Highways Regulations (which were of below average quality). ESs for large scale develop-ments were usually better than those for smaller projects, and there was some correlation bet-ween length of the ES and quality. A more recent study for the ESRC by the University of Manchester used 48 cases, and showed a great improvement in the quality of ES after 1990, although a subsequent decline in quality was indicated after 1992. Overall, only just over half of the ESs were regarded as satisfactory (Wood and Jones, 1995).

Aggregated assessments of quality are also provided from other perspectives, although often with less rigorous analysis. In their 1991 study, Wood and Jones noted that LPAs tended to be more generous than the EIA Centre in their evaluation of ES quality. An analysis by the IEA of 15 ESs produced during 1990-91 suggested that 65% were broadly satisfactory, although Coles and Tarling (1992) noted that the ESs were not selected randomly with some submitted voluntarily by consultants for review.

Objectivity appears to be a general problem with the EA systems, both in the UK and elsewhere. Lee and Brown (1992) identified particular problems of bias, with 60% of ESs unsatisfactory. Ginger and Mohai (1993) noted that lack of objectivity in ESs is a major issue under the US National Environmental Policy Act system. However a more recent study by Nelson (1995) based on 36 planning authorities in the UK suggests both an improvement in ES quality since the early years and a possible improvement in objectivity, with only 20% of statutory consultees considering ESs inadequate in terms of credibility and accuracy of findings.

2.4.3 Disaggregated approach

Disaggregated approaches focus on quality with regard to particular types of projects, to content/treatment of individual topic areas in the ES (e.g. landscape, flora and fauna), to performance in various stages of the EA process (e.g. description of baseline conditions, consideration of alternatives), and to presenta-tion and other aspects of the ES. There is little comparable time series data.

Lee and Dancey (1993) provided a disaggre-gated analysis of ES quality across the four areas of the Lee and Colley review criteria, for 83 ESs across all regulations. ESs were shown to be least satisfactory in: the identification and evaluation of key impacts and most satisfactory in their description of the development, local environment and baseline conditions. Other studies have examined ESs produced for parti-cular categories of development. For example, a survey of 22 ESs for extractive industry projects using the Lee and Colley criteria found that their quality was lower than that of other ESs produced in the period 1988-1991, but that 'there is evidence of considerable improvement in quality between the initial period of EA implementation (1988-1989) and the subsequent period (1990-1991)' (Kobus and Lee, 1993).

2.5 Determinants of ES/EA quality

The literature directly identifies several possible factors which could affect the quality of ES/EA. These fall broadly under two headings: those relating to (i) the type and size of project, and (ii) the experience of the developer, consultant and competent authority. The literature also raises, in a more indirect manner, (iii) other factors, including EA related guidance and factors related to the planning process.

2.5.1 Factors related to the project

Several factors related to the project can affect ES/EA quality: Annex I or Annex II; size of the project; type of regulation; and type of project. One could thus expect EA/ES quality to improve if: more Annex I projects are proposed; more large projects are proposed; more projects are proposed that fall under the Electricity and Pipeline Works (Assessment of Environmental Effects) Regulations 1989 or the Environmental Assessment (Scotland) Regulations 1988; and proportionately more projects normally associated with good ESs are prepared. Of course, such quality determinants may not be easily subject to policy influence.

2.5.2 Factors related to the experience of the developer, consultant and competent authority

<u>ES prepared in-house or by outside consultant</u>
The study of 83 ESs prepared between mid-1988 and early 1991 (Lee and Brown, 1992) showed that ESs prepared in-house by the developer have generally been of much poorer quality than those prepared by outside consultants. A review of 47 ESs (including 22 ESs for extractive industry projects) also showed that ESs prepared by consultants were generally of higher quality (36% satisfactory) than those prepared by the developer (18%) (Lee et al., 1994). Another study of 70 ESs prepared before 1993, which focused on their visual assessments, concluded that while some developers - particularly for extraction and public utility projects - prepare

high quality assessments, other in-house ESs were of very poor quality (Mills, 1994).

Experience of the developer and consultant
ESs prepared by developers or consultants with previous experience of EA are generally of better quality than those prepared by an inexperienced developer or consultant (Lee and Brown, 1992). The case study of 22 ESs for extractive industry projects by Kobus and Lee (1993) showed 43% of ESs were satisfactory where the developer had previous experience with EA, compared with 14% where the developer had no previous experience. Skehan (1993) also notes an 'observed correlation' between ES quality and the experience of consultants and planning authorities. However the experience of the producer is not, of itself, a guarantee of quality (Lee and Dancey, 1993). The Ministry of the Environment for Luxembourg, where ESs for major projects tend to be carried out by large and experienced German, French and Swiss consultancies, noted that ESs contained several deficiencies, especially with regard to the lack of detailed examination of alternatives, and in forecasting of possible effects (CEC, 1992).

Experience of the competent authority
A study of the quality of 40 Irish ESs (Lee and Dancey, 1993) suggests that better ESs are prepared for competent authorities with more experience than those with less experience: 59% of ESs received by competent authorities which had dealt with 10+ ESs were satisfactory, compared with 25% for competent authorities with less than 10 ESs. The experience of UK LPAs is increasing. A study by Oxford Brookes University's Impacts Assessment Unit (as yet unpublished) of more than 2,000 ESs prepared by late 1994 revealed that only one county council and less than 20% of district and borough councils had not yet received any ESs.

2.5.3 Other possible factors
The literature suggests other possible factors which could influence the quality of ES/EAs.

Presence of EA-related guidance and legislation
Over time, increasing guidance has emerged concerning EA directly or indirectly. This guidance may be improving ESs/EAs. It is also possible that, although the Lee and Colley (1990) and IEA (1991) review criteria may only be in limited use in local planning authorities, they may be used by the environmental consultants and developers preparing ESs to ensure that their ESs are adequate. Books on EA, such as those of Glasson et al (1994), Petts and Eduljee (1994) and Wathern (1988), may also be used as guidance by all the parties involved in EA.

The presence or absence of guidance may also be a reason for the difference of ES/EA quality between different types of projects. The early guidance on EA for planning-related projects given in the DoE's (1989a) booklet *Environmental Assessment: a guide to the procedures* was more comprehensive than that for highway schemes (DoT Standard HD 18/88), forestry projects (the Forestry Authority's booklet *Environmental Appraisal of Afforestation Projects)* or marine salmon farms (no guidance). It is also noticeable, although as yet undocumented, that the DoT's ESs have improved enormously since the publication of the *Design Manual for Roads and Bridges Volume 11: Environmental Appraisal* in 1993. The greater amount of information about some types of projects may improve the quality of their ESs. For instance, the monitoring studies that have been carried out, independently of the EA process, regarding restoration of opencast coal sites, may lead to improved EAs for future opencast projects. Similarly, increasing experience with relatively new types of projects such as incinerators, motorway service stations or windfarms, may improve their ESs over time.

Better ESs may also be prepared for local authorities that have prepared handbooks on EA - for instance Kent, Cheshire and Essex county councils - than for those that have no such handbook. Finally, better ESs may be prepared in areas where the development plan is more precise, an environmental audit of the local authority has been carried out, or an environmental appraisal of the development plan has been conducted.

Factors related to the project planning process

The stage in project planning at which the development application and ES are submitted is likely to affect the quality of the ES. The better a project is defined, the more detailed the ES predictions can be. On the other hand the EA process is unlikely to be effective if an ES is merely produced after all the relevant decisions have been made. In parallel, the longer the timescale of project planning, the more likely it is that adequate studies can be carried out for inclusion in the ES/EA.

Proposed projects that are likely to involve a high-profile environmental issue, or that are likely to go to public inquiry, may result in a better-quality ES than those without such issues. Similarly, high-quality ESs may result where there are significant financial implications for the agency responsible for the project if problems should occur. The resources, especially the financial resources, available to the developer are also likely to affect ES quality. However, the way in which these resources are made available may be just as crucial. If the need for EA is only considered near the end of project planning, or if the EA is given low priority despite adequate funding, then the ES/EA quality is likely to suffer.

Other issues related to interactions between the parties involved in EA

MacCallum (1987) raises some additional issues regarding the commitment of the various parties to the EA process, and the communication between these parties. The quality of an ES is likely to reflect whether the agencies and people responsible for the proposed project are committed to the EA process. This commitment will affect the resources allocated to the EA, the stage in project planning where EA is carried out, and the general tone of the resulting ES. Greater levels of consultation with statutory consultees and the public are also likely to affect ES quality by ensuring that the ES covers all relevant issues and adequately considers alternatives and mitigation measures. In this sense, it is likely that ES quality will be linked to the experience, interpersonal skills, mutual expectations and co-operation of the people involved in the EA process.

CHAPTER THREE

Methodology for the review of the matched pairs of Environmental Statements

3.1 Introduction

Changes, if any, in ES quality between the pre-1991 and post-1991 periods were assessed at two levels. In the 'macro study' the research team assessed levels of quality and changes in quality for 25 matched pairs of ESs using a number of separate but comparable review frameworks. In the 'micro study' for a sub-set of 10 matched pairs, the assessment of quality was pursued from various perspectives, and case files and interviews were used to assess quality for whom. The 'micro study' considered the EA process. It also provided a vehicle to undertake the parallel research on identifying the costs (and benefits) associated with the production of ESs, and developing a methodology for assessing such costs.

3.2 Choice of 25 matched pairs of ESs

3.2.1 Criteria for choice

The 25 matched pairs of ESs were chosen from those ESs submitted under the EA Planning Regulations, where a final decision had been taken on the planning application. Each pair includes one statement from the 1988-1990 period and one from the 1992-1994 period. ESs submitted in 1991 were excluded (with minor exceptions noted below) to provide a clear break between the two sets. Within this framework, the following criteria were used to identify the 25 pairs:

- they should be representative of the full population of projects for which ESs have been submitted, with reference to:
  - Schedule 1 and 2 projects
  - type of project (e.g. waste disposal, road
  - geographical area (i.e. England, Scotland and Wales; and broad regions of England);
- wherever possible, the specific type of project, county/Scottish region and type of developer should be the same for both projects in each pair. It was possible to match project types and areas well; matching types of developers was more difficult, although some useful pairings were achieved; and
- the ESs should be available. For this reason a 'reserve set' of 50 ESs was also drawn up, structured in the same way as for the 'original set'. Projects from the 'reserve set' were only used when it proved impossible to obtain the ESs for the 'original set'.

3.2.2 Chosen set of 25 matched pairs of ESs

The details of the 50 ESs are included at Appendix 4 structured by the following 12 characteristics:

- project number - for this research
- type of project
- location/description
- date of submission

Figure 3.1: Location of matched pairs of ESs used in the study.

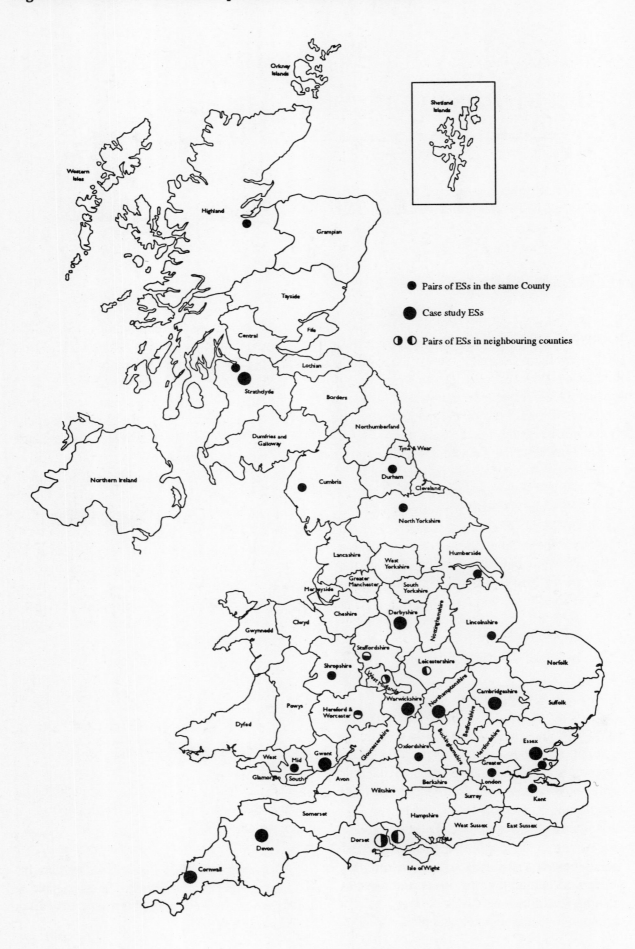

- schedule (EA Regulations)
- whether the ES is in the Oxford Brookes University collection
- whether the ES is in the Department of the Environment Library
- county/Scottish region
- local planning authority
- developer/agent
- whether the application went to appeal
- final decision on the application.

The types of projects used in the analysis are summarised in Table 3.1, and the geographical location of the matched pairs is illustrated in Figure 3.1.

Table 3.1: 25 matched pairs of ESs - project types.

• road	4
• waste disposal-landfill	8
• sewage treatment works	2
• waste treatment/recycling	2
• extraction-sand and gravel	4
• opencast coal	4
• industrial-heavy	2
• industrial-light	2
• new settlement	2
• mixed use development	4
• power station	2
• wind farm	4
• leisure	2
• agriculture	2
• food processing	2
• reservoir	2
• motorway service area	2

3.2.3 Representativeness of the matched pairs

It is estimated that approximately 1700 ESs were submitted under the EA Planning Regulations between July 1988 and September 1994 (Frost and Frankish, 1995). As such the 25 pairs constitute only about 3% of the total population. Comparison of Figures 3.2(a) and 3.2(b) show a very good level of fit between the 25 pairs and the total population, by project sector. But although the pairs provide a representative sample for aggregate analysis, several of the sub-groups (by project type, or geographical area for example) are too small for meaningful disaggregated analysis.

Figure 3.2: Representativeness of the full population of the 25 matched pairs (EA Planning Regulations: September 1994).

(a) Population.

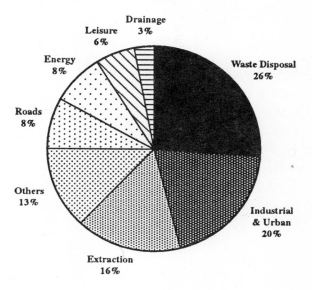

(b) Matched Pairs

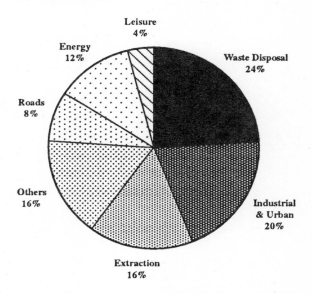

The geographical distribution of the matched pairs in Figure 3.1 reflects 13% (three pairs) of ESs submitted under the Scottish EA Planning Regulations; together with the 9% (two pairs) submitted in Wales. Figure 3.1 shows a broad spread of pairs across the regions of mainland UK. For other dimensions of comparison, it is estimated that within the total population approximately 8% of ESs have been submitted by local authorities themselves; the figure for the sample is 12%. The known (minimum) number of ES applications that have gone to appeal is 6%; the number of appeal cases within the sample is 8%.

3.2.4 Limitations of the pairs

The specifications for the matched pairs are numerous and exacting. The research achieved a 'good fit' of pairs to specification, but there are bound to be a few limitations although these are not considered to affect the represen-tativeness of the findings.

On receiving some of the ESs it was found that the proposed projects were not exactly as expected, and this has led to a small number (4 out of 25) of slightly imperfect matches. This problem is partly a consequence of the widespread confusion over which Schedule classes projects fall under. A further minor limitation relates to geographical proximity. It was decided at the outset that the pairs should consist of ESs as geographically close together as possible. This is because the quality of ESs may vary between different areas, perhaps reflecting factors such as the level of impor-tance attached to environmental considerations and the networking nature of EA activity. Most of the pairs are made up of ESs submitted within the same county or Scottish region, however in three cases ESs in neighbouring counties had to be used. Finally, two of the ESs were submitted in early 1991, one of them associated with an appeal.

3.2.5 Obtaining ESs

Overall, acquiring the ESs proved to be an unexpectedly lengthy process. The DoE Library was helpful in providing copies where these had reached the Library. The Institute of Environmental Assessment also provided an efficient loan service. The best source for the remainder of the ESs proved to be the local planning authorities, but only where spare copies were available. Encouragingly LPAs were mindful not to release file copies on the grounds that the ESs should always be available for public inspection. In some cases, LPA officers were cautious of providing spare copies without the consent of the applicants. In many instances surprise was expressed that the DoE could not provide the ESs, as copies had been forwarded to the DoE regional office. Consultancies were generally a poor source of ESs. In most cases they were very wary of providing ESs for this research without the consent of clients. There was widespread concern that published research findings should not enable individual cases to be identified.

Despite these difficulties, there were very few cases where it was just impossible to obtain a copy of the ES, or where the information used to draw the sample was found to be incorrect. In a small number of cases, parts of the ES documentation were missing. These were followed up and adjustments made to the relevant assessments as appropriate.

3.3 Approach to the review of the 25 matched pairs

3.3.1 Review frameworks

Each ES has been reviewed against the criteria in a comprehensive framework, detailed in Appendix 5. The framework has eight sections:

1 description of the development;
2 description of the environment;
3 scoping, consultation and impact identification
4 prediction and evaluation of impacts;
5 alternatives;
6 mitigation and monitoring;
7 non-technical summary; and
8 organisation and presentation of information.

The framework has been organised so as to allow the review of ESs in four ways, according to how they meet:

1 the 'regulatory requirements', often referred to as the 'minimum requirements' required by the Town and Country Planning (Assessment of Environmental Effects) Regulations (1988), as outlined in *Environmental Assessment: A Guide to the Procedures* (DoE, 1989), and amplified in the recent report on *Good Practice on the Evaluation of Environmental Information for Planning Projects* (DoE, 1994d). The 'regulatory requirements' criteria in Para 2 to Schedule 3 of the TCP(AEE) regulations were amplified as nine more detailed criteria, which are indicated by the asterisked items in Appendix 5.
2 a comprehensive checklist devised by the Impacts Assessment Unit (IAU) research

team, and based on the review criteria in Glasson, Therivel and Chadwick (1994). This includes all the elements in Appendix 5.

3 the Lee and Colley review package (1991), which assembles the various criteria into a tiered framework with four assessment areas: description of development, local environment and baseline conditions; identification and evaluation of key impacts; treatment of alternatives and mitigation; and communication and presentation of results.

4 the EU Review Checklist (CEC, 1993), developed for DG XI of the EC by Environmental Resources Management as a method to review the completeness and suitability of environmental information from a technical and decision making viewpoint. The criteria are organised in eight review areas: description of the project, outline of alternatives, description of the environment, description of mitigation measures, description of effects, non-technical summary, difficulties compiling information, and general approach.

The focus in this research is on the amplified 'regulatory requirements' reflecting a simple, legislative baseline of information, and the IAU checklist reflecting a more comprehensive and sensitive approach. Limited use of the Lee and Colley package allows comparison with earlier studies; use of the EU checklist provides a further methodological comparison of wider European relevance.

Three scoring systems are used in the quality review:

- for 'regulatory requirements':
  √ - the ES covers the topic satisfactorily
  X - the ES does not cover the topic satisfactorily
  nr - topic not relevant to the project
- for the IAU checklist, and for the Lee and Colley package:
  A - generally well performed, no important tasks left incomplete
  B - generally satisfactory and complete, only minor omissions and inaccuracies
  C - can be considered just satisfactory despite omissions and/or inadequacies
  D - parts are well attempted but must, as a whole, be considered just unsatisfactory because of omissions and/or inadequacies
  E - not satisfactory, significant omissions or inadequacies
  F - very unsatisfactory, important task(s) poorly done or not attempted
  nr - not relevant
- for wthe EU checklist:
  c - complete: all information relevant to the decision making process is available
  a - acceptable: information presented is not complete, however omissions need not prevent the decision-making process proceeding
  i - the information presented contains major omissions; additional information is necessary before the decision making process can proceed
  nr - information not relevant to specific context of project

3.3.2 Review process and issues

Each of the 50 ESs in the 25 matched pairs was reviewed twice, by separate reviewers, giving a total of 100 individual reviews on completion of the exercise. The individual reviews were all carried out by the five members of the research team. The ESs were allocated randomly across the team members to avoid any 'regular pairings'. The approach also avoided, in almost all cases, a reviewer taking both ESs in any one pair, as the review of the first might influence the second. Each reviewer of an ES reviewed it 'blind', before negotiating agreed summary 'scores' with the paired reviewer, against the 'regulatory requirements', IAU checklist, EU checklist, and Lee and Colley package.

The review process was piloted with a small number of ESs to identify any relevant issues and to achieve a consistent approach. In addition to problems of partial documentation in a few cases, issues raised included, *inter alia*, the nature of 'alternatives', the nature of an ES 'non-technical summary', and when the 'minimum requirements' were fulfilled. A consistent approach to the issues raised was agreed through the piloting process and through regular team meetings during the review exercise.

3.3.3 Approach to the assessment of the findings

The findings of the research have been analysed on both aggregated and disaggregated bases. The aggregated approach assesses the overall change in the quality of ESs over the study period. The disaggregated approach seeks to identify subsets of change, for example for particular parts of the EA process. The assessment of changes in quality includes several dimensions: level of quality, consistency of quality and extremes of quality, amongst others. A quantitative approach to the assessment of change is used wherever possible. For the 'regulatory requirements', the assessment considers the number of ESs meeting *all* the requirements, the numbers meeting *specific proportions* of the requirements, and changes in *particular strengths* and *weaknesses* of the ESs in relation to the requirements.

The assessments against the IAU checklist and the Lee and Colley package use the A-F scoring system. Following Lee and Colley, categories A, B and C are regarded as 'satisfactory' and D, E, F are regarded as 'unsatisfactory'. An alternative approach packages A and B as 'good', E and F as 'poor', with C and D as 'marginal'. It is also possible to identify changes in consistency of quality against the eight IAU and four Lee and Colley review areas, over the study period. What constitutes a 'significant' change in quality partly depends on perspective. However some changes may be more significant than others, for example from a D to a C in contrast with a change from an F to an E. An improvement in the spread of quality within an ES, with a more consistent approach, rather than one based on a few high spots, might also be regarded as a significant change.

3.4 Approach to the detailed review of 10 matched pairs

3.4.1 Choice of the 10 matched pairs

Similar criteria to those discussed earlier were also applied to the identification of cases for the 'micro study'. With a smaller sample, the fit with the wider population is more approximate. Table 3.2 lists the project types and Figure 3.3

Table 3.2: 10 detailed matched pairs of ESs-project types

• road	Essex
• waste disposal-landfill extension	Warwickshire
• sewage treatment works	Devon
• extraction-sand and gravel	Northamptonshire
• opencast coal	Strathclyde
• new settlement	Cambridgeshire
• mixed use development	Gwent
• windfarm	Cornwall
• leisure	Derbyshire
• reservoir	Hampshire/Dorset

shows the breakdown of the 20 projects by category. A comparison with Figure 3.2(a) shows a good fit still with the total population.

3.4.2 Review process

The review process involved two stages. The first involved getting together as much background information about the processing of the relevant application as possible. It was considered important to minimise the number of visits to LPAs and to use publicly available file material to establish the main actors and quality issues raised in the individual cases. The following background papers were acquired from the LPAs for each of the 20 cases:

Figure 3.3: Representativeness of the 10 matched pairs

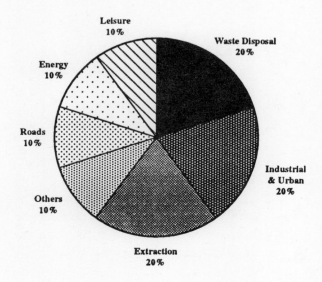

- relevant committee reports and minutes
- the responses of statutory consultees, local councils (including parish and community) and non statutory bodies (including local and national pressure or interest groups) to consultation on the application; and
- the Council's decision letter and, where the project went to appeal, a copy of the final decision letter.

The papers provided the basis for the assessment of 'quality for whom' issues and of the effectiveness of the EA process as a whole. The second stage of the research involved the structuring of the assessment of the issues. A questionnaire (Appendix 7) provided a framework. Section 1, on the ES in the application process and quality for whom issues was completed from the papers as far as possible. Section 2 focused on the various participants' views of the quality of the ES and the EA process, and also on the costs and benefits of EA. This involved telephone and/or face to face interviews.

CHAPTER FOUR

Changes in quality: findings from review of 25 matched pairs

4.1 Introduction

This chapter discusses the results of the analysis of the 25 matched pairs of ESs. The ESs' coverage of the simple 'regulatory requirements' are discussed first, then their coverage of the more comprehensive IAU criteria. Appendix 6 discusses the ESs' quality in terms of two other review systems' criteria: those of Lee and Colley, and of the EU. In each case, the pre-1991 results are presented first, then the post-1991 results; finally the two sets of results are compared.

4.2 Simple 'regulatory requirements'

4.2.1 Pre-1991

Table 4.1 summarises how the pre-1991 ESs covered the 'regulatory requirements'. Almost two-thirds of these ESs did not fulfil *all* the requirements. If a less restrictive approach is taken, where uncertain coverage (denoted as 'marginal') is taken to mean coverage, then:

- the proposed development (criterion 1) is adequately described in 19 (76%) ESs,
- the data needed to identify and assess the development's main environmental impacts (criteria 2 and 3) are adequately covered in 18 (72%) ESs,
- the proposed development's likely impacts (criterion 4) are considered adequately in 16 (64%) ESs, but only 15 (60%) of these consider all of the listed impacts (criterion 5),
- mitigation measures (criteria 6 and 7) are adequately covered in 17 (68%) ESs, and
- a non-technical summary is included in 16 (64%) ESs, but only 14 (56%) of them cover the main findings of the ES.

Under the less restrictive approach, 10 (36%) ESs covered all 9 criteria adequately, 6 (24%) covered 5-8 criteria, and 9 (36%) covered 4 or less criteria. A more rigorous approach would lower some of these numbers slightly.

4.2.2 Post-1991

Table 4.1 also summarises how the post-1991 ESs covered the requirements. Eleven (44%) of the 25 ESs fulfilled all the requirements, and 14 (56%) did not. Again, with a less restrictive approach:

- the proposed development is adequately described in 21 (84%) ESs,
- the data needed to identify and assess the development's main environmental impacts are adequately covered in 23 (92%) ESs,
- the proposed development's likely impacts are considered adequately in 20 (80%) ESs, but only 19 (76%) consider all of the listed impacts,

Table 4.1: Simple 'regulatory requirements' (25 pairs of ESs)

Criterion	pre-1991 ESs			post-1991 ESs		
	covered	marginal	not covered	covered	marginal	not covered
1. Describes the proposed development, including its design and size or scale.	19		6	18	3	4
2. Defines the land areas taken up by the development site and any associated works, snd shows their location on a map.	19		6	22	1	2
3 Describes the uses to which this land will be put, and demarcates the land use areas.	17	1	7	23		2
4. Considers direct and indirect effects of the project, and any consequential development.	15	1	9	19	1	5
5. Investigates these impacts in so far as they affect human beings, flora, fauna, soil water, air, climate, landscape, intersections between the above, material assets, and cultural heritage.	14	1	10	16	3	6
6. Considers the mitigation of all significant negative impacts.	17		8	21	2	2
7. Mitigation measures considered include modification of the project, the replacement of facilities, and the creation of new resources.	15	2	8	20	3	2
8. There is a non-technical summary which contains at least a brief description of the project and environment, the main mitigation measures, and a description of any remaining impacts.	16		9	19	1	5
9. The summary presents the main findings of the assessment and covers all the main issues raised.	13	1	11	15	3	7
All criteria	9 (36%)		16 (64%)	11 (44%)		14 (56%)

- mitigation measures are adequately covered in 23 (92%) ESs, and
- a non-technical summary is included in 20 (80%) ESs, but only 18 (72%) of them cover the main findings of the ES.

Under the less restrictive approach, 12 (48%) ESs covered all 9 criteria, 11 (44%) covered 5-8 criteria, and 2 (8%) covered four or less criteria.

4.2.3 Features of inadequate ESs

These include inadequately described projects where, for example, the ES did not include a map of the proposed location and site, did not make a clear distinction between the proposed development and other interlinked developments, did not provide a clear indication of the development's likely physical presence, transport implications, or level of employment, gave inconsistent information, and relied on the planning application to give details of the project.

Figure 4.1: Coverage of simple 'regulatory requirements': pre-1991 v. post-1991(25 pairs of ESs)

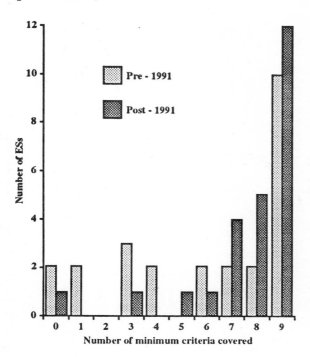

Inadequate environmental data primarily took the form of minimal or non-existent environmental baseline information. The likely impacts of the proposed development were often described, but not in a comprehensive or well-structured manner. Examples of poor impact prediction included an ES which only considered landscape, and an ES for waste disposal which did not discuss flora, fauna, landscape, material assets, or the cultural heritage. In some of these cases, the impacts may have been minor, but no explanation was given for why the impacts were not addressed in the ES.

Examples of inadequately discussed mitigation measures include ESs which described the project so badly that it was impossible to determine the mitigation measures proposed, and ESs which did not address the likely impact and thus did not propose mitigation measures for it. Several ESs did not contain a non-technical summary; some of the ESs themselves were at best inadequate summaries. In other cases a summary had been prepared, but was of very poor quality or else the main body of the ES was called a non-technical summary.

4.2.4 Changes over time

When considering whether all the nine criteria had been covered, of the 25 pairs of ESs, 11 (44%) showed no change, 8 (32%) showed an improvement (from not covering all the criteria to covering them all), and 6 (24%) became worse over time. Those ESs which cover all the main criteria increased from 36% before 1991 to 44% since 1991, and those ESs which cover 6 or more of the 9 criteria rose from 64% before 1991 to 92% since 1991. Of all the 225 possible criteria (25 ESs x 9 criteria each) which could have been fulfilled, 145 (65%) were fulfilled pre-1991 and 173 (77%) post-1991. A more detailed analysis shows that there has been a consistent shift towards a fuller coverage of the 'regulatory requirements' (see Figure 4.1). Coverage for each of the criteria has improved, from an average of about two-thirds pre-1991 to more than 80% post-1991. The non-technical summary is now the worst-covered criterion (approximately 75% coverage).

Overall the level and quality of coverage of the ES 'regulatory requirements' is improving, although more than half of recent ESs still do not fulfil all of the criteria used in this study.

4.3 Oxford Brookes University Impact Assessment Unit (IAU) criteria

4.3.1 Pre-1991

Table 4.2 summarises how the pre-1991 ESs covered the IAU criteria. The overall quality was just unsatisfactory (D), with 9 (36%) poor ESs (D/E to F), 12 (48%) marginal ESs (C to D), and 4 (16%) good ESs (A to B/C). Using an alternative marking system, 9 (36%) of the ESs were satisfactory (A to C), 13 (52%) were unsatisfactory (D to F), and 3 (12%) were in between (C/D). Those sections of the ESs that were carried out best (C/D) were the organisation and presentation of information, description of the development and the environment, and mitigation and monitoring. Scoping and consultation, impact identification and prediction, and the non-technical summary were generally carried out just unsatisfactorily (D). Alternatives were generally not considered or considered very badly (E).

4.3.2 Post-1991

Table 4.3 summarises how the post-1991 ESs covered the IAU criteria. Overall quality was satisfactory (C), with 5 (20%) poor ESs, 11 (44%) marginal ESs, and 9 (36%) good ESs. Alternatively, 15 (60%) of the ESs were satisfactory, 9 (36%) were unsatisfactory, and 1 (4%) was in between. The sections that were carried out to the highest standard (C) were the organisation and presentation of information, the description of the development and the environment, and impact prediction and evaluation. The remaining sections were generally carried out to a standard between satisfactory and unsatisfactory (C/D), with the exception of the consideration of alternatives, which was just unsatisfactory (D).

4.3.3 Features of good and poor quality ESs

Good quality ESs included, *inter alia*, clear structure, good coverage of complex issues, commitment to monitoring, good consultation, and good coverage of alternatives. Poor quality ESs included, *inter alia*, low objectivity and complex technical content, and had the charac-teristics of a supporting document to an application rather than an ES in its own right.

4.3.4 Changes over time

As shown in Figure 4.2, based on the IAU criteria, in general there seems to have been a significant improvement in the quality of ESs between 1988-90 and 1992-94, from an average of just unsatisfactory (D) to just satisfactory (C). The proportion of poor ESs (D/E to F) dropped from 36% to 20%, and the proportion of good ESs (A to B/C) more than doubled from 16% to 36%. The proportion of satisfactory (A to C) ESs almost doubled, from 36% to 60%, and the proportion of unsatis-factory (D to F) ESs dropped from 52% to 36%. The range of quality, however, has remained unchanged: from B to F.

Statistical analysis was undertaken to check whether the difference in the pre-1991 and post-1991 mean quality of ESs could have occurred by chance because of the limited size of the sample (i.e. 2 x 25 cases from a much larger population). The so called 't-test' is often used to establish this. A more suitable statistical test in this case is the Wilcoxon matched pairs test, as the two samples (early and later ESs) are structured on the basis of matched pairs. The Wilcoxon test is also more appropriate than the t-test because it is a non-parametric test rather than a classical one. This means that it does not require the assumption that the population is normally distributed about the mean quality score. Application of this test showed that the difference in the mean quality scores between early and later samples of ESs is significant at the 0.05 level (i.e. the probability of these data occurring by chance is less than one in 20).

Quality has also improved in each of the eight main categories of assessment. The ESs' description of monitoring and mitigation improved only marginally, but the other categories generally improved by about half of a mark (eg. from D to C/D). A particular improvement was seen in the approach to alternatives, which was on average not satisfactory (E) before 1991, and improved by more than a mark to just unsatisfactory (D) since 1991. This can be seen as an important change for a category which is not mandatory under UK regulations.

Figure 4.3 shows the amount of change over time within each pair. Of the 25 pairs, the review showed an improvement in 15 pairs, and a reduction in quality in 9 pairs. In general the pairs of ESs improved by 1 or 1.5 marks (eg. from D to C, or D to B/C) over time. However in two cases the ESs improved by 3.5 marks (eg. from E/F to B), and in another the ESs improved by 4 marks (from F to B). In 7 of the 9 pairs where quality deteriorated over time, it was only by 0.5 or 1 mark. However for one pair quality dropped by 1.5 marks, and for another it dropped by 2.5 marks.

Overall, using the IAU criteria, the quality of ESs rose from just unsatisfactory pre-1991 to just satisfactory post-1991. The ES sections carried out to the highest standards are the description of the development and environment, prediction and evaluation of impacts, and organisation of presentation of findings. Coverage of alternatives -- not a minimum requirement -- was poor pre-1991, but has

Table 4.2: IAU criteria: pre-1991 ESs (25 ESs).

Criterion Quality	A	A/B	B	B/C	C	C/D	D	D/E	E	E/F	F	
1. Description of the devel.		2	3	2	6	3	2	2	2	2	1	C/D
2. Description of the env.		1	4	2	8	1	2		4	2	1	C/D
3. Scoping, consultation, & impact identification		1	1	1	6	3	4	3	4	1	1	D
4. Prediction & evaluation of impacts			3	2	6	1	4	3	4		2	D
5. Alternatives			1	1	1	2	3	1	7	I	8	E
6. Mitigation & monitoring		1	5		6	1	6	3	3			C/D
7. Non-technical summary		1	5	2	3	1	3	3	2		5	D
8. Organisation & presentation of information		1	5	2	4	3	4	3	3			C/D
Overall ES			3	1	5	3	4	1	4	1	3	D
		good = 4 (16%)			marginal = 12 (48%)			poor = 9 (36%)				25 (100%)
		satisfactory = 9 (36%)				3 (12%)		unsatisfactory = 13 (52%)				25 (100%)

Table 4.3: IAU criteria: post-1991 ESs (25 ESs).

Criterion	A	A/B	B	B/C	C	C/D	D	D/E	E	E/F	F	total
1. Description of the devel.			8	4	5	1	3	1	1	1	1	C
2. Description of the env.		1	6	3	9		5		1			C
3. Scoping, consultation, & impact identification			4	2	6	4	5		3	1		C/D
4. Prediction & evaluation of impacts		1	8	1	3	3	5		3	1		C
5. Alternatives		1	4	1	6		3		7	2	1	D
6. Mitigation & monitoring			6	1	6	2	4	3	2	1		C/D
7. Non-technical summary		1	6	3	4	2	3	1	1	1	3	C/D
8. Organisation & presentation of information			8		8	3	4		1	1		C
Overall ES			5	4	6	1	4	3	1		1	C
		good = 9 (36%)			marginal = 11 (44%)			poor = 5 (20%)				25 (100%)
		satisfactory = 15 (60%)				1 (4%)		unsatisfactory = 9 (36%)				25 (100%)

improved significantly. However, although 15 pairs of ESs showed improvement over time, 9 became worse. More than one-third of post-1991 ESs were still unsatisfactory, and 20% were poor.

4.4 Other review criteria: Lee and Colley, and EU

The application of the Lee and Colley review criteria, and the more recent EU review criteria, to the 25 matched pairs, is detailed in Appendix 6 to reduce repetition in the main text. The findings complement very closely those presented in this chapter.

The application of the Lee and Colley criteria show an overall improvement in quality, from D to C, over the two periods. The distribution of the assessment of quality for the pre-1991 period, with approximately 33% satisfactory and 67% unsatisfactory ESs, mirrors quite closely the earlier findings by Wood and Jones (DoE, 1991a), and Lee, Walsh and Reader (1994). The shift in the distribution of quality over the two periods from 10 satisfactory, 14 unsatisfactory plus 1 marginal, to 14 satisfactory, 7 unsatisfactory plus 4 marginal also complements the findings using the other review methods.

Figure 4.2: Marks for IAU criteria: pre-1991 v. post-1991.

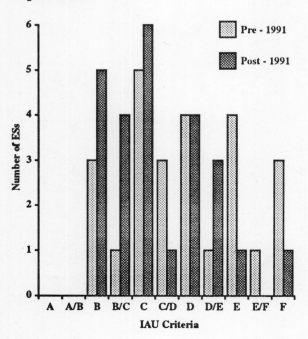

Figure 4.3: Change in ES quality (IAU criteria) within pairs: pre-1991 v. post-1991.

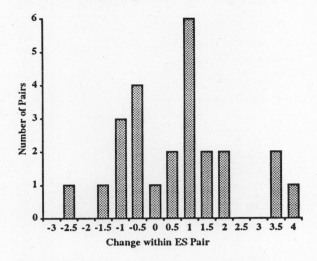

The EU review framework uses a more limited range of assessment (ES complete, acceptable or incomplete). Again there is a positive shift from 4 complete, 10 acceptable and 11 incomplete (pre-1991) to 7 complete, 12 acceptable and 6 incomplete (post-1991).

4.5 Conclusions for findings from 25 matched pairs

There has been an improvement in the quality of ESs over this quite short period whichever of the criteria are used.

In terms of the simple 'regulatory requirements', 44% of the post-1991 ESs fulfilled all the criteria, compared with 36% of pre-1991 ESs. A more detailed analysis reveals that 92% of the post-1991 ESs fulfilled 6 or more of the 9 criteria, compared with 64% of the pre-1991 ESs.

In terms of the criteria established by the Impacts Assessment Unit, the overall quality of ESs rose from just unsatisfactory pre-1991, to just satisfactory post-1991. This can be seen as an important shift across a key threshold, although the overall post-1991 grade of C still signifies omissions and inadequacies in many ESs. The percentage of satisfactory ESs has increased from 36% to 60%. The difference between the ESs' lower level of fulfilment of the simple 'regulatory requirements' criteria and the higher level of achievement of satisfactory quality according to the IAU criteria can be attributed in part to the rigorous demands of meeting certain criteria in the simple checklist (esp. the list of impacts in criterion 5), and of the yes/no scoring system, compared with the grading system used in the IAU approach.

Yet the analysis of the matched pairs provides some evidence of 'two steps forward, one stepback'. Under the IAU criteria, although 15 pairs of ESs showed improvement over time, 9 became worse. Similarly there are 16 better and 9 worse using Lee and Colley criteria; 12 better and 6 worse using EU criteria. As such, whilst the overall conclusion is one of improvement in ES quality over the two periods, it must be noted that the quality of many ESs is still not good.

CHAPTER FIVE

Perceptions of ES quality in the planning application process: findings from the case studies of 10 matched pairs

5.1 Introduction

For the participants in the planning application process the quality of environmental statements is largely in the eye of the beholder. Each participant has his/her own perspective on the merits or otherwise of a project and this will inevitably colour the assessment of the ES. Furthermore each participant, or group, will want, expect, demand to find different types of information and different levels of detail within an ES. In an assessment of the changing quality of ESs it is important to both recognise these different requirements and to discover whether or not an ES is meeting the needs of those who participate in the EA process. As such the approach in this chapter differs from the previous quantitative approach of the independent researcher; the focus is on a more perceptual approach to ES quality, illustrating the various perspectives of the participants in the EA process.

To make this assessment 42 individuals who had participated in the processing of the 10 case study matched pairs were interviewed. From the background papers supplied by local planning authorities, the planning case officer, the developer/consultant and the key third party participants in the EA process were identified for each case. It was not possible to interview all the individuals identified because some had moved on, some were on leave or otherwise unavailable and some could no longer remember the case in sufficient detail to provide answers. In a few cases individuals declined to participate.

The interviewees contacted for the study were 5 developers/-consultants, 15 planning officers and 22 third parties (statutory and non-statutory consultees, individual objectors, interest groups).

From these interviews it is possible to provide a general picture of the perceptions held by the participants in the EA process on both the use and quality of ESs in that process. Section 5.2 examines the overall quality of the ESs from the perspectives of the participants set against the assessment grades given by the IAU in this study. The section also outlines the different perceptions held by planning officers and consultees on whether or not an ES meets the minimum regulatory requirements. In section 5.3 the perceptions of the participants on the quality and use of ESs in the planning process are discussed. The key issue for this study, the change in the quality of ESs over time, is discussed in Section 5.4 from various perspectives. Section 5.5 deals with the interviewees' views on the costs and benefits of the EA process. In section 5.6 the main themes to emerge from the study of the ten matched pairs are discussed.

5.2 Quality of case study ESs

Interviewees were asked to give their own personal assessment of the ESs and to grade them 'good', 'marginal' and 'poor'. These assessments are set out in Tables 5.1 and 5.2 with IAU grades which have been translated thus: A - B/C = good; C - D = marginal; and D/E - F = poor.

Table 5.1: Overall assessment of ESs: pre 1991 cases

Case No	Planner(s)	Consultee(s)[2]	IAU
1	N/I[1]	marginal	marginal
3	marginal	poor	poor
5	good	poor	marginal
7	N/I	N/I	marginal
9	marginal	N/I	marginal
11	N/I	marginal	marginal
13	poor	poor	marginal
15	marginal	poor	marginal
17	marginal	poor	poor
19	good	N/I	good

Table 5.2: Overall assessment of ESs: post 1991 cases

Case No	Planner	Consultee(s)[2]	IAU
2	good	good	good
4	marginal	poor	good
6	good	marginal	good
8	N/I	poor	marginal
10	good	N/I	marginal
12	N/I	good	poor
14	marginal	poor	marginal
16	marginal	marginal	marginal
18	marginal	poor	poor
20	good	marginal	good

Note 1: N/I No interview
Note 2: Where more than one consultee has been interviewed the assessment recorded in the above tables is an aggregate of the grades given by all consultees.

For the pre-1991 case studies none of the interviewed planning officers had used any structured review system and only one of the consultees had used any formal review criteria and that was an internal checklist. In the post-1991 cases, only one planning officer had used a structured review system (Lee and Colley) and one of the consultees had used an internal review system. Tables 5.1 and 5.2 illustrate that the perception of practitioners, mainly without the aid of any formal review system, differs from that of the more objective and systematic review by the IAU. In only two out of the 12 cases with responses from both planner and consultee, was there general agreement over the quality of the statement. Approximately half of the interviewed officers agreed with the IAU assessment and there was 31% agreement between consultees and the IAU. There was agreement between planning officers and consultees in 25% of cases, with consultees consistently rating the ESs lower than did the planners.

Tables 5.1 and 5.2 show also that planning officers considered the ESs to be generally of better quality than had the IAU whereas consultees thought the ESs to be generally of poorer quality than had the IAU. The assessment of the developer/consultant was not included in this analysis because of the low number of respondents and the fact that, perhaps naturally, they tended to consider their work to be good; although some, with the benefit of hindsight, were self critical.

Where a consultee had given a 'poor' grade this tended to be due to the inadequacy of the section of the ES which dealt with their particular field of interest. For example, those statutory and non-statutory bodies with some form of wildlife conservation or archaeological interest raised particular concerns about the quality of the ESs in their representations on the planning applications. In only two of the 20 cases were the ecology issues considered by all consultees to have been handled satisfactorily. In one case local amenity bodies complained that the ES concentrated far too much on the strategic

Table 5.3: Perceptions of meeting minimum regulatory requirements: pre-1991.

Minimum regulatory requirements	yes	no	d/k
Planners	86%	14%	0%
Consultees	11%	55%	34%

Table 5.4: Perceptions of meeting minimum regulatory requirements: post-1991.

Minimum regulatory requirements	yes	no	d/k
Planners	100%	0%	0%
Consultees	50%	16%	34%

issues of need and policy and failed to identify the key issues that concerned them, such as noise and traffic generation. The main reason given by planning officers for a 'poor' or 'marginal' assessment was inadequacy of information.

Tables 5.3 and 5.4 record the percentage of interviewees who considered that the ES they had examined met the minimum regulatory requirements. The planners' perceptions of ES compliance with minimum requirements are higher than those of the independent researchers (see Chapter 4), reinforcing the findings of earlier research (Wood and Jones, 1991). When asked why an ES had failed to meet the requirements 26% of respondents referred to the inadequacy of information. The other main reasons given were lack of early consultation or adequate scoping (26%); and conciseness of presentation (22%). In both the pre- and post-1991 samples there was a relatively high percentage of consultees who could not judge whether or not an ES met the required standard.

5.3 A semi-quantitative analysis of the ES/EA participants and the planning application process

The ES is submitted with a planning application as part of the planning process for major projects. The ES is produced by the developer as part of the EA process; it forms part of the environmental information required by the 1988 Planning EA Regulations to be taken into consideration before a decision can be made on an application to which the Regulations apply. The ES is therefore an important, if not central, component of the planning application where EA is carried out and *it has a role within the whole process from pre-application stage to the final decision and beyond.*

5.3.1 Perspectives on ES in the planning application process
Pre-Application Stage
Because of the largely discretionary system for determining whether EA is required for a Schedule 2 project, it is not uncommon for a planning authority to require the submission of an ES after the lodging of the planning application and this was the case in almost half of the 20 cases. Overall 58% of planning officers said they were involved in some way with the ES or the development proposal prior to submission, while only 14% of the consultees interviewed said they were involved with the EA prior to being consulted by the local planning authority.

Negotiations and consultation at the pre-application stage are generally considered by interviewees to be important, not just for the quality of the ES but for the processing of the application as a whole. However, early consultation does not necessarily mean the consultees will be satisfied with the outcome. In some cases there had been prior consultation with bodies which then lodged objections to the planning application on the grounds of insufficient assessment of impacts within their field of interest. Archaeological bodies considered it necessary for developers to carry out far more on-site investigations before producing the ES as statements contain "little more than is available from county records". There clearly is a difference between information collection and consultation over matters such as scoping, prediction and mitigation. Local wildlife trusts considered it important that developers and their consultants make use of local knowledge and welcomed the approach that kept them involved "before, during and after". With some major projects, such as large urban development schemes or sewage treatment schemes which have many separate elements, officers have considered it important to involve the Council members at the earliest possible stage so that they are aware of how the design of the development is progressing and are involved at a stage when changes can still be made.

Post Submission

During the processing of the application officers generally saw little difference between projects subject to ES and any other applications of similar complexity and controversy. Once the application is lodged "the development control process takes over". There appeared to be general agreement among planning officers that receiving environmental information in one document, at an early stage, was helpful and probably saved them time. One officer said "When the system first appeared I was rather sceptical because I believed we had always taken all these matters into account. Now I am a big fan of the process. It enables me to focus on the detail of individual aspects at an early stage".

For the review of the statements local authorities appear to rely chiefly on the statutory and non statutory consultees to assess the different elements of the ES. In none of the case studies had the Council used an external consultancy to carry out a review; although two had subsequently done so with other ESs. Where there is insufficient information on ecological or archaeological issues within the ES, and where the site or species under investigation is not designated for statutory protection, it appears more normal for the local planning authority to put the developer and the consultee in direct contact rather than formally request further information under the terms of the Regulations. It was felt by two consultees interviewed that while this was useful and did help to redress some deficiencies in the ES they, particularly the smaller bodies, 'lacked the muscle' to demand too much.

An objector to one planning application felt that the ES was far too technical even for the planning officers who, in the objector's view, just ignored the deficiencies as they were under pressure to grant planning permission no matter what the impacts. An ecological consultant who prepared part of a statement for a developer, emphasised the need for co-opera-tion between the producers and consumers of ESs during the processing of the application: "some local authorities keep their cards very close to their chests, which makes them much more difficult to work with".

Members are normally made aware of the ES through the Committee agenda notification and copies of the ES tend to be made available to members by depositing a copy in the Members' Room prior to the meeting. Local ward members, those most affected by the proposal, occasionally asked for personal copies of the ES, but on the whole planning officers did not believe that members looked at an ES before the application is due to be determined, "if at all".

Decision making

Three out of 15 planning officers did not consider the ES to have had much influence on the planning decision and the quality of the ES was not considered to be a major issue in decision making. Three planning officers and one developer/consultant believed the reason for this was that not all issues discussed in ESs are material planning considerations and the weight to be attributed to any consideration was a matter of planning judgement.

Four officers felt that Members were not generally interested in the ES and preferred to rely on officer's reports for a summary of the issues. One officer reported that "by and large they are happy to rely on our reports rather than wade through the whole thing themselves". One officer suggested that members were more influenced by their site visit than by the ES which they rarely looked at. In some of the cases discussed the officers felt that because of the nature of the project, particularly major

Table 5.5: Extent to which ES influenced planning decision.

Extent of influence (no. of respondents) /participants	Much	Some	Little/ None	D/K
Planners	3	9	3	O
Consultees	1	7	6	8
Developers/ consultants	1	2	1	1

infrastructure schemes, they did not have a great deal of leeway to either refuse or even seek changes to the scheme. The quality of an ES was considered to be important to decision making where it provided information on the need for a project and where it helped the officer form a recommendation by addressing the main objections to the project.

Many interviewees from non-statutory bodies felt excluded from the decision making process and one national non-statutory wildlife body complained that if the Nature Conservancy Council (NCC) or the Countryside Commission did not object then their own objections went largely ignored by the developer and the local planning authority; this despite their often far more detailed knowledge about local conditions. One officer with a statutory nature conservation body commented "It is better not to object to a development, because if we object then we do not even get planning conditions put on the development. We are better off with conditions".

Post-Decision
Once the decision has been made planning officers are only involved where there is need to check the imposition of conditions or the terms of Section 106 Obligations or if the project has been refused and is to be the subject of appeal. One planning officer also pointed out that many aspects of major developments do not require planning permission, particularly where statutory undertakers are involved. Post decision changes can be made to a project which have their own impacts and over which the planning authority has little or no control. In one such case the project changed virtually beyond recognition after planning permission had been granted. Because of problems unforeseen in the ES, changes had to be made which resulted in major impacts not considered by the ES. In another case an ES failed to recognise that a waterway affected by a development was in the process of being colonised by otters, a fact which only came to light after the decision had been made by the Council. Here an ecological impact assessment was required through the S106 in an attempt to afford the otters some last minute protection. The ecological assessment ended up discounting the need to protect otters, but it identified badger setts, a pond with great crested newts, and bat species on the site.

5.4 Perspectives on changing quality of ESs

5.4.1 'Quality for whom' criteria
Interviewees were asked how they would assess the ESs against what can be termed the 'quality for whom' criteria of 'comprehensiveness', 'objectivity' and 'clarity'. The results, which are set out in Tables 5.6 and 5.7, do indicate an improvement over time. For comprehensiveness 31% of interviewees considered the pre 1991 cases as being good and 38% as poor. In the post-1991 cases 55% of interviewees considered

Table 5.6: Quality of ESs: pre 1991.

Criteria/grade	Good	Marginal	Poor
Comprehensiveness	31%	31%	38%
Objectivity	18%	37%	45%
Clarity of information	25%	56%	19%

Table 5.7: Quality of ESs: post 1991.

Criteria/grade	Good	Marginal	Poor
Comprehensiveness	55%	27%	18%
Objectivity	41%	35%	24%
Clarity of information	55%	38%	7%

the cases to be good and 18% poor. This same trend was true of the other two criteria.

These figures represent the views of participants and while in most cases the individual ESs are being assessed more than once, each individual interviewee was giving an independent assessment and was unaware of the assessment provided by others. This means that, for example, of all those interviewed from the pre-1991 cases 45% considered the ES *they had seen* to be of poor quality in terms of its objectivity, while 24% of the post-1991 interviewees considered the ES *they had seen* to be poor in terms of objectivity. *The results do, therefore, indicate that in terms of issues such as objectivity there has been a perceived improvement in the quality of ESs.*

5.4.2 Quality for planning officers

One very experienced planning officer believed that the reputation of the consultants producing the ES influenced the view of the LPA on the quality of the ES and the reliability of the information provided. Many planning officers felt that the use of experienced and reputable consultants was the best way of achieving good quality ESs. Just over 40% of planning officers agreed that there had been an improvement in the overall quality of ESs over time. Most of the remaining officers felt this was difficult to assess when individual officers see so few ESs and the ones they do see tend to be for different types of projects which raise different issues. Where officers did consider there had been an improvement it was seen as only marginal, with a lack of adequate scoping recognised as still being a major problem. They also noted that "alternatives is the weakest aspect of all the ESs we have dealt with".

ESs were still seen as being targeted at gaining planning permission and minimising the implications of impacts. One officer said that ESs are getting better "but getting bigger" and that a "big ES is still a rather daunting prospect". There is some evidence that more local authorities are making use of systematic review systems such as the DoE's checklist in *Evaluation of Environmental Information for Planning Projects: a Good Practice Guide* (1994e) or the services of the IEA.

5.4.3 Quality for consultees

Some of the consultees felt unable to make detailed assessments of the quality of the ESs as they normally (one interviewee thought in 90% of cases) only received one section of the ES. One interviewee who deals with more than 5 ESs per year could see no discernable trend to suggest ESs were improving and noted that "we still receive some pretty atrocious statements". Another interviewee believed that there remained great variation, and that individual environmental consultants were still capable of producing both good and bad ESs. One national nature conservation body felt that the single biggest omission was a sufficiently detailed assessment of secondary impacts and that this was a "major problem with ESs".

A senior staff member of a local nature conservation trust who deals with between 6 and 12 ESs per year - including those on which they act as consultant - did have a "general impression that things were improving". This was seen as being largely because local planning authorities have more experience in handling ESs and because there is more pre-submission consultation. Objectivity and clarity of presentation were considered to be important, improving and yet still wanting. A senior officer with a national statutory consultee commented "good consultants lead to good development. It is vital to have good environmental consultants".

5.4.4 Quality for developers/consultants

One developer considered the overall quality of the ES in question to be 'good' on the basis that "it achieved planning permission". A consultant who had helped to produce an ES in 1989 considered it to have been 'good' "given knowledge about EIA at the time", implying that it would not be given such a high grade if it were produced today. There was also a view from another consultant that developers are increasingly recognising the need for environmental protection and are prepared to bring in consultants at the very beginning so that a development could be designed around those needs. Another view was that "more guidance and experience has led to more constructive criticism from consultees". Other consultants believed that in general quality was improving by virtue of the experience and the higher expectations for ESs of pressure groups in particular.

5.5 Costs and Benefits

Early resistance to the imposition of EA into the UK was based on two main propositions: that it would cause additional expense and delay in the planning process; and that the development control process already took into consideration the matters that would be considered in an ES. This review, and particularly the case study approach, has facilitated a return to those two, perhaps linked, concerns. The evidence from Sections 5.1, 5.2 and 5.3 suggests that planning officers and developers see merits in the system and accept that EA does permit much earlier and more comprehensive assessment of impacts than under standard development control procedures. This Section examines this further and relates it to the financial and other costs of producing, processing and reviewing ESs.

5.5.1 Planning Officers

None of the planning officers could provide anything more than a guesstimate of costs and time spent on reviewing ESs and felt that dealing with the ES and the planning application were one and the same and "just part of the job". Estimates for reviewing the ES and associated consultation ranged from five hours to 6-8 months of staff time. Although many officers found it difficult to separate the work generated by the ES from the work on the planning application, the general view was that considerably more time was involved in an application with an ES than one without. Planning officers handling ES cases tend to be development control team leaders and above, and therefore staff costs would generally be higher than with standard planning applications.

Four officers felt that the EA process had brought time savings in that the ES provided information on the project at an early stage and (ideally) in a single comprehensive document. This saved the officer, and internal consultees like environmental health officers, from having to collect the information themselves. One officer took the view that "projects are generally improved as a result of EA" and did not believe that the process slowed down consideration of the application and that the "sixteen weeks was far more realistic than the eight normally given for consideration of applications".

Most planning officers agreed that projects and the environment benefited generally as a result of EA and the submission of an ES. One officer said that the most important role of an ES is to "flag up problems early and direct them to the right people". However another officer thought that the benefits really only come when an ES is of good quality and that "it can speed things up when there is no need to ask for further information", and another considered the ES merely acted as a PR exercise for the developer, as the information provided would have been required in any case by the LPA. A `step in the right direction' was the view of one officer and this may be an appropriate LPA conclusion.

5.5.2 Consultees

There was some agreement that the introduction of EA had created a more structured approach to handling planning applications and that in considering applications "we at least have something to work from rather than have to dig around for information ourselves". However when issues had been `scoped out' of the assessment, the relevant consultee was left in the same position as with non-EA applications in that they had to carry out the assessment themselves. An officer of a statutory body was of the view that "some of my objections are not because the impacts are bad but because I have not been given any information on impacts or any explanation of why (field of concern) has been left out of the assessment". One interviewee from a local nature conservation trust considered a major benefit of the system was that it gave them data on sites that they would not have been able to afford to collect themselves. Many consultees found it difficult to attribute any environmental protection or improvements to the quality of projects to the EA process as they could not judge what the outcome would have been without EA.

The benefits of the system were linked to the quality of the ES by a number of consultees. One experienced consultee was of the view that "good ESs save costs for consultees" and saw good quality ESs as having good, clearly set out summaries, logical presentation of information and a good rationale behind the approach. The key area was seen as the description of the baseline environment. One non-statutory consultee saw the EA process as going some way towards reaching common ground between the parties involved, in an otherwise too often confrontational planning system. As with the planning officers there were, *at the time of ESs reviewed,* no structured time recording or accounting systems for assessing the cost of reviewing ESs as part of the consultation process, although many respondents said that *now* such an accounting system is in place. One respondent felt much more precise costs to be largely irrelevant as "that is what we are here for", although another interviewee from a similar organisation stated that "we do need to prioritise which develop-ments we get involved in because of time - a large development can take up a week of officer time".

Estimates of staff time spent by statutory consultees ranged from 4 hours to 1.5 days (with a financial estimate of £200 for senior officers and clerical assistance for a total of 10 person-hours); and from 1 hour to 2 weeks for non-statutory consultees. Longer periods were estimated by internal consultees within the Local Authority. Statutory and non-statutory consultees both felt that the heavy involvement in assessing and commenting on ESs at the peak of activity in the early years of the Regulations had markedly affected their work. In some cases, this had led to new units being set up (for instance, co-ordinating ecological and geological advice, on both external and internally prepared ESs); in others, it had led to a restriction of the consultee's involvement in priority sites or issues.

5.5.3 Developers and consultants

Consultants and developers are more likely to have detailed cost data, related to their contractual arrangements, although they may, for reasons of client confidentiality, be unprepared to release the information. Consultants use their internal accounting systems to keep close records of ES costs, including sub-contracting costs. However it is still sometimes difficult to separate out the ES cost when a consultant begins work on a project at pre-design stage and follows it right through to a decision following a public inquiry. Estimates of the costs of ES preparation varied from 22 person-days, at a cost of £5000, to 3-4 person months with, in addition, further work contracted out. A senior member of staff with a major environmental consultancy took the view that "the ES does not necessarily slow down the application. The more organised approach makes it more efficient and in some cases it allows issues to be picked up earlier. The ES can thus speed up the system". Another was of the view that "it has shortened the planning application stage but has lengthened the period before the ES is submitted".

5.6 Themes and issues

The developers/consultants, planning officers and third party consultees interviewed in this study cannot be claimed to be a statistically representative sample of the many thousands of individuals who participate in the planning application process. However they, and the background file papers supplied by local authorities, do provide an insight into the use of ESs in the planning application process and the importance of the quality of an ES to that process. From the evidence reported in this Chapter a number of themes can be identified which reinforce those which have emerged from the main study of the 25 matched pairs:

(1) there is a perception, from several perspectives, that there has been an improvement in ES quality over the two study periods. This relates to several criteria, including compliance with minimum regulations, comprehensiveness, objectivity and clarity of information;

(2) whilst perceptions on the level of ES quality vary, there is agreement that early consultation, scoping and adequacy of information are central to meeting the needs of all participants, to producing good quality ESs and to the time taken in considering the applications;

(3) good quality ESs help to speed up the application process;
(4) the use of reputable consultants is likely to lead to the production of a good quality ES; and
(5) there is little quantitative assessment made of the costs or time taken in processing planning applications accompanied by an ES; however there is some evidence that good quality, comprehensive, statements can bring resource savings.

CHAPTER SIX

Determinants of ES quality and changes in quality

6.1 Introduction

Chapter 2 identified a number of possible factors which could affect the quality of ESs and the EA process, including: factors related to the project, factors related to the nature and experience of the various participants in the EA process (especially developer, consultant, local authority, statutory and non-statutory consultees), and other factors, including EA related guidance and legislation. The explana-tory value of these factors is now discussed in relation to the findings in both Chapters 4 and 5, with a focus first on level of quality, and then on changes in quality.

6.2 Determinants of ES quality

6.2.1 Factors related to the project

Previous studies have suggested that certain *types of projects* have been associated with better quality ESs and with a positive shift in quality (Therivel et al, 1992; Kobus and Lee, 1993; and Hinson, 1994). Some projects, often Schedule 1, may be particularly high profile and attract substantial attention and resources. Our sample of 50 ESs cannot be disaggregated to reveal such variations, with any statistical validity, because of the small numbers of ESs reviewed for the various project types. However, there is some indication that, under the Planning (AEE) Regulations, better quality ESs are associated with developments such as windfarms, (more recent) waste disposal and treatment plants, sand and gravel extraction schemes and opencast coal; whereas generally poorer quality ESs are associated with mixed use developments, new settlements, leisure proposals and agricultural schemes. The sub-sample of Schedule 1 projects is too small to draw any meaningful conclusions for this category of projects.

Figure 6.1 provides an indicative correlation of the relationship between *size of project* and quality of ES. Projects have been grouped into three size categories (small, medium and large) according to their size in relation to the indicative size thresholds for determining whether EA should be required. A small project has been taken to be one which is less than 75% of the threshold size; a medium sized one between 75% and 125%; and a large one over 125%. Most of the project types in the sample have quantified thresholds as set out in Figure 6.1. However for the last four types, a less standardized approach was adopted as no indicative size threshold exists. The graph shows that if all the ESs including and above a 'C' on the IAU criteria assessment are classed as satisfactory, and those below as unsatisfactory, then in broad terms, *the larger the project the more satisfactory the ES tends to be*.

Figure 6.1: Relationship between 'size' of project and quality of ES.

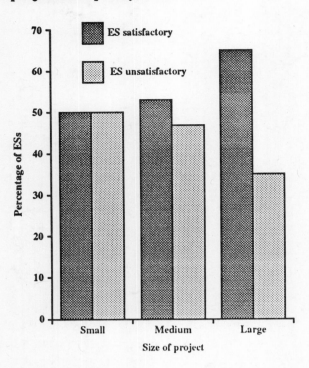

referred to in Chapter 2, about the influence of the *nature and experience of the developer, consultant and competent authority* on the quality of the ES.

Table 6.1 shows that ESs produced in-house by developers are on average of much poorer quality than those produced by outside consultants. Table 6.2 shows a similar but less marked, difference in quality between ESs produced by an independent applicant in comparison with those produced by the decision maker (local authority).

Figure 6.2 clearly demonstrates the significance of ES experience of both the consultant and the local authority as a determinant of ES quality. Whereas approximately only 50% of consultants with little or no prior experience produce satisfactory ESs, most of those with experience of eight or more ESs produce satisfactory

Project type	Size thresholds
Turkey farms	50,000 birds
Opencast coal	50 ha
Sand and gravel	50 ha
Industrial plants	25 ha
Industrial estates	20 ha
Urban developments	5 ha, 10,000 m² shops, offices or commercial uses
Local roads	10 km
Other infrastructure	100 ha
Waste disposal	75,000 tonnes pa
Windfarms	10 wtgs, 5 MW
Motorway service areas	5 ha
Sewage treatment works	Catchment population
Special waste disposal	Threshold as for non hazardous waste (75,000 tonnes pa)
Minor power stations	MW output compared to windfarm threshold (5 MW)
Reservoir schemes	Other infrastructure threshold (100 ha)
	Extent of extraction vis sand & gravel projects (50 ha)

6.2.2 Factors related to the nature and experience of the various participants in the EA process

The findings from the review of the 50 ESs reinforce many of the findings of other studies, statements. Other studies (Weston, 1995) show that consultants with more experience are much more likely to produce the whole ES rather than just sections, in contrast with consultants with less experience. This may aid quality control and bring more resources to bear on the EA process. There is a similar correlation between (review) experience and ES quality for local authorities, but with slightly lower levels of satisfactory quality. With a few exceptions, the review experience of county councils far exceeds that of district councils.

6.2.3 Other factors

Other determinants of ES/EA quality include EA guidance and legislation, aspects of the project planning process, and issues related to the interactions between the parties involved in the EA process. Changes in government and general *guidance* (e.g. from academic publications) are discussed further in section 6.3. There is also local guidance for some areas, and there is limited evidence from the study of 50 ESs that those produced for local authorities with their own 'customised' handbooks on EA are of a higher than average quality.

The *stage in project planning* at which the development application and EA are submitted is likely to affect the quality of the ES. The better a project is defined, the more detailed the ES predictions can be. ESs can be submitted for projects, as part of an *outline* or more normally, a *detailed planning application*. Another indicative finding of the study is that the quality of ESs for outline applications is significantly poorer than that for detailed applications. At the outline stage, the project is not so clearly defined, so predictions about the project's likely impacts are necessarily less accurate and mitigation measures are less fully considered. The average quality mark for these projects, using the IAU criteria, is D/E, which is somewhat below the general average. In addition, the average quality of these projects does not appear to be improving over time.

Other determinants of ES quality can be the *commitment of the project proponents* to the EA process, the levels of consultation with statutory consultees and the communications between the various parties involved in project planning. These determinants may be reflected in the *resources allocated* to the EA process. They cannot be ascertained easily through the ES review process discussed in Chapter 4. However the more detailed investigation of particular case studies discussed in Chapter 5

Table 6.1: ES quality: ESs prepared by consultants or in-house developer - IAU criteria.

Grade	Consultant	In-house	Both
A			
A/B			
B	6	1	1
B/C	5		
C	7	2	2
C/D	3	1	
D	2	5	
D/E	4	1	
E	2	3	
E/F		1	
F	2	2	
Average	C/D	D/E	B/C

Table 6.2: ES quality: ESs prepared by the decision-maker or independent applicant - IAU criteria.

	decision-maker (eg. county council transport or minerals department)	someone else
A		
A/B		
B		8
B/C	2	3
C	1	10
C/D		4
D	1	6
D/E	1	4
E	2	3
E/F		1
F	2	2
Average	D/E	C/D

Figure 6.2: ES quality: consultant and local authority experience (50ESs).

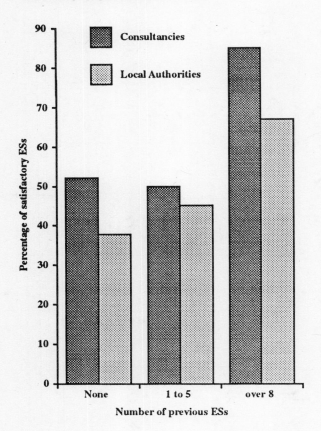

indicates that LPAs, consultees and developers are making in some cases substantial resource inputs to the EA process.

The length of the ES is sometimes seen as a proxy for commitment and resources, and previous studies have shown some correlation between *ES length* and ES quality (Lee and Brown, 1992). Table 6.3 compares the quality of the ESs with their length. The results show a general improvement with increased length, from an average of E/F for ESs of less than 20 pages, to an average of C for those of more than 150 pages. As ESs become much longer than 150 pages, however, quality becomes more variable; although the longer ESs may contain more information, their length may also be a symptom of poor organisation and coordination. ESs seem to be becoming longer over time, but the variation seems to be more related to project type than to time.

Spearman's Rank Correlation test was used to test the strength of this apparent correlation between ES length and quality and the coefficient was found to be +0.54. This test was used

Table 6.3: ES quality v. length of ES (no of pages).

	<10	10-19	20-49	50-99	100-149	≥150	massive
A							
A/B							
B				2	3	1	2
B/C				3	1		1
C			1	4	1	3	2
C/D			4		1		
D			2	1	2		1
D/E			2	2		1	
E	1	2	2				
E/F				1			
F		2	1				
Average	E	E/F	D	C	C	C	C

because it does not require the assumption that the distribution of the population is normal; it is likely to be skewed in terms of both quality and length. Application of the test did show that the Spearman's Rank Correlation coefficient was significant at the 0.05 level. In other words there is a less than 1 in 20 probability that this positive correlation between ES quality and length occurred by chance.

6.3 Explanation of changes in ES quality

Evidence for explanations of changes in ES quality can be found in trends in the factors outlined in the previous section.

6.3.1 The project mix

The mix of the projects for which ESs are prepared may change over time. If the change is towards projects normally associated with good ESs and away from those associated with poor ESs there will be an improvement in the quality of the population of new ESs. There is some evidence of such a shift with the number of ESs for waste treatment plants, opencast coal extraction and windfarms increasing steadily, and those for business parks and new settlements decreasing in proportion (Therivel et al, 1992, 1993; Frost et al, 1994, 1995). By taking matched pairs of project types the analysis of this study has reduced the significance of this factor. But project related factors also include the impact of the refinement and increasing familiarity of more recent developments. Some types of projects are relatively new; for instance, windpower has only recently gained momentum in the UK, with an increase in the installed energy from 10 MW in 1991 to 131 MW in 1993. Lack of familiarity with the projects and uncertainty over the applicability of the Regulations in the

early days led to a minimalist approach by the developers and LPAs, but it is claimed that 'since these early days the wind industry has dramatically improved its expertise in the EA process, and now works with major consultants on the full range of environmental issues. Substantial 3-volume ESs are now the 'norm', with particular emphasis on landscape, noise, ecology and planning policy' (Hinson, 1994).

6.3.2 Experience of the EA process

The experience of the parties involved in the EA process may change. Higher quality is associated with more experienced developers, consultants and LPAs. Experience has increased over time, as exemplified in Table 6.4 which compares the ES experience of developers, consultants and LPAs involved in the pre-1991 and post-1991 sets of ESs used in the matched pairs for this study.

However, the number of previous ESs prepared does not necessarily equate to experience. New consultancies may employ experienced practitioners; consortiums and partnerships may form, just for one project. Experience may also lead to complacency and an overstandardised approach. There may be some evidence of EA tasks now being delegated to more junior members of staff, as the novelty value wears off.

Another factor mentioned several times by respondents in the case studies was the importance of the 'good' environmental consultant. This refers to a variety of factors. There may be a growing influence of 'standard bearing' consultants producing 'top drawer ESs'. Legal advice to developers may be a force for good in the use of good consultants. The differential accreditation by bodies such as the IEA may also force up standards. There are also more ESs in the public domain to provide benchmarks. These are open to scrutiny, not least from fellow practitioners; although, as noted, ES availability is not good.

There is also some conflicting evidence on the use of consultants. The macro study suggested a marked shift away from ESs prepared in-house and towards those produced by consultants. On the other hand the ENDS (1995) report suggests that major developers are increasingly employing their own in-house specialists, with consultancies only being brought in as troubleshooters when an innovative approach is required. Returns on the investment in ES quality are increasingly being realised by

Table 6.4: Changes in the ES experience of developers, consultants and LPAs in the pre-1991 and post-1991 periods.

	Pre-1991 set of ESs	Post-1991 set of ESs
Average number of ESs produced by/for the developer[1]	1	3
Average number of ESs produced by the consultant[1]	1	4.5
Average number of ESs reviewed by the LPA[1]	3	8

[1] At the date of submission of the ES used in the study.

developers, as they no longer pay for the 'learning curve' of EA, as consultancies become more competitive, and as data becomes more available.

Pre-application consultation on the scope of the ES, between the LPA and the developer/consul-tant, seems to be increasing. Early consultation and scoping have been highlighted as very important for the quality of ES for all participants. Indeed it can be argued that one of the most valuable roles of the EA process is to encourage such consultation. In addition, some development plan policies now refer directly to issues expected to be covered in particular types of ESs (e.g. waste and mineral local plans). Consultees, statutory and non-statutory also appear to be making more use of the ES in the planning application process - although they may each have a very limited perspective on the ES, focusing narrowly on particular topics of concern, and providing little advice to the LPA on the overall quality of the ES, if in fact they see the whole ES.

6.3.3 Other factors, including guidance and training

There has been an improvement in relevant guidance for the preparation of ESs, although much of this has been quite recent and may have had little or no impact on the ESs reviewed in this study. For much of the post-1988 period the main guidance has been through the Official Circulars (DoE, 1988a) and the DoE's (1989a) booklet *Environmental assessment: a guide to the procedures*. However, it is likely that the impact of the latter was more limited in the early rather than later period. The 'micro-study' of 20 cases also suggested that the use of standard ES review frameworks, such as Lee and Colley, by LPAs is limited. Other guidance is much more recent, including the latest DoE *Good Practice guides* on *preparing environmental statements for planning projects* (1994g) and on the *evaluation of environmental information for planning projects* (1994e). There has also been a major growth since the late 1980s in EA training support, both in the form of courses (MSc, short courses etc.) and in relevant literature (books, articles etc.) and this is most likely to have increased the EA expertise in developers, consultants and LPAs in the post-1991 period.

It appears likely that over time EA activity has become better integrated into the various stages of the planning and development process, and that there is increasing familiarity with the role and potential of ESs from a wide range of parties, including pressure groups and the statutory consultees. The expectations of ESs of all parties, including pressure groups, are rising over time, and this is a force for improved quality. It is also likely that those reviewing ESs are demanding higher standards in 1995 than in 1991, influenced themselves by guidance and examples of good practice.

CHAPTER SEVEN

Conclusions and recommendations

7.1 Conclusions

7.1.1 Context of conclusions

The literature review, informed by the survey of key practitioners, indicated that a focus on ES quality should consider both 'minimum requirements' for such statements, as specified by legislation, and also 'best practice'. Key quality criteria for ESs should include compliance with regulations, comprehen-siveness of information, adequacy of methodology, clarity and organisation of information, effective communication and accessibility to relevant audiences, and transparency, objectivity and impartiality. In addition ES quality cannot be divorced from the wider context of EA quality, and the EA and planning application process. Quality is also partly in the 'eye of the beholder', and another emphasis in this study has been on 'quality for whom' including, in particular, the perspectives of the local planning officer, the statutory and non-statutory consultee, and the developer and consultant. All these aspects have been included in this study.

7.1.2 Changes in ES quality: pre-1991 to post-1991
Findings from the 'macro-study' of 25 matched pairs

There has been an improvement in the quality of ESs, over this quite short period of time, whichever of the criteria are used. In terms of the simple 'regulatory requirements', 44% of the post-1991 ESs fulfilled all the criteria, compared with 36% of the pre-1991 ESs. A more detailed analysis reveals that 92% of the post-1991 ESs fulfilled six or more of the nine criteria, compared with 64% of the pre-1991 ESs. However, notwithstanding this improvement, more than half of recent ESs still do not fulfil all of the criteria and this is a cause for some concern.

Using the criteria established by the Impact Assessment Unit (IAU), the overall quality of ESs rose from just unsatisfactory pre-1991, to just satisfactory post-1991. This can be seen as a significant shift across a key threshold, although the overall post-1991 grade of C still signifies omissions and inadequacies in many ESs. The ES sections carried out to the highest standards are the description of the development and environment, prediction and evaluation of impacts, and organisation and presentation of findings. Coverage of alternatives - not a minimum requirement - was poor pre-1991, but has improved significantly. However, the analysis of the 25 matched pairs provides some evidence of 'two steps forward, one step back', with 15 pairs of ESs showing improvement over time but 9 becoming worse. Similarly there are 16 better and 9 worse using the Lee and Colley criteria, and 12 better and 6 worse using the EU criteria. Overall, using the IAU criteria, more than one-third of post-1991 ESs were still unsatisfactory, and 20% were poor, this reinforces the cause for concern noted above.

Findings from the 'micro-study' of 10 matched pairs

The analysis of the files for the 20 case studies, plus interviews with 42 individuals who had participated in the processing of the various cases, provided an insight into the use of ESs in the planning application process and the importance of the quality of an ES to that process. A synthesis of the information from these sources illustrates the differing perceptions of ES and EA quality. In general, planning officers considered the ESs to be of better quality than the IAU research team whereas consultees thought them to be worse that had the IAU team. The statutory and non-statutory bodies which raised most concerns about the quality of the ESs were those with some form of wildlife conservation or archaeological interest. However all parties agreed that quality, in relation to meeting 'minimum requirements', had improved pre-1991 to post-1991.

There was a clear view from all parties that pre-application activities, with early consultation, negotiation and participation in the scoping of the ES, were important to the quality of the ES and to the time taken in considering the application. The studies highlighted the reliance of the LPAs on the consultees to assess the various elements of the ES post-submission. There were mixed views about the influence of the ES in decision making. Only 5 out of 42 interviewees thought the ES had 'much' influence on the decision, but another 18 thought it had 'some' influence. The greatest level of scepticism was from the consultees.

The 'comprehensiveness' and 'clarity of information' of ESs was assessed by the various respondents to have improved over the study period. There was also a perception of some improvement in the objectivity of ESs, although two thirds of respondents for the post-1991 ESs still regarded the objectivity as either marginal or poor. Local planning officers perceived the lack of adequate scoping as still being a major problem in the EA process. Consultees emphasized the role of good consultants in producing good quality ESs.

Little quantitative assessment was made by LPAs of the costs taken in processing planning applications accompanied by an ES. Evidence from the case studies gives mixed views, with some officers perceiving time savings and others perceiving additional costs. Most agreed that projects and the environment benefited generally as a result of EA and submission of an ES; there was also some evidence that good quality, comprehensive, statements speed up the process and bring resource savings for LPAs and consultees. Consultees also welcomed the more structured approach to handling planning applications associated with the introduction of EA. Of all the participants in the process, consultants and developers have the most detailed cost data, although there can still be problems in separating out ES specific costs. There was again a view from this group that a good ES could speed up the application process.

7.1.3 Determinants of changes in ES quality: pre-1991 to post-1991

The study has identified a number of possible factors which could affect the quality of ESs and EA including factors related to the project, the experience of the various participants in the EA process, and EA related guidance and legislation.

Certain types of project have been associated with better quality ESs. The mix of the projects for which ESs are prepared may change over time. Whilst not particularly subject to policy influence, a change towards projects normally associated with good ESs and away from those associated with poor ESs, plus the increasing familiarity with more recent project types, may lead to an improvement in the quality of the population of ESs.

High quality is also associated with more experienced developers, consultants, consultees and LPAs. Experience has increased for all of the main participants in the process, although it may be quite limited still for many practitioners. Higher quality has been associated in particular with the 'good' environmental consultant; although there is concern about the dangers of complacency, and overstandardisation perhaps as a result of price competition. There is also some concern about the opportunistic new entrant to the field with little to offer other than a low price. Early consultation and scoping have been highlighted as very important for the quality of the ES for all participants; pre-application consultation appears to be increasing although it is still very limited, or none existent, in many cases. Overall, the expectations from all parties of ESs, including pressure groups, are rising over time and this is a force for improved quality.

Improvements in guidance and training may also explain some of the improvements in quality. The DoE's (1989) booklet *environmental assessment: a guide to the Procedures* has been particularly useful. This has subsequently been supplemented with further DoE guidance (1994e, 1994g), by review frameworks, guidance from the IEA and various books and articles; although much of this subsequent material may have been too recent for many of even the post-1991 case studies. There are also more ESs in the public domain to provide evidence of good practice.

7.2 Recommendations

7.2.1 Context of recommendations

The study has revealed some improvements in the quality of ESs from those produced pre-1991 to those produced post-1991. It has also revealed that a substantial proportion of ESs are not of satisfactory quality, and that within the content of the ES documentation itself and within the wider EA procedures, a number of problem areas can be identified. The earlier study for the DoE (1991) on the implementation of the EA Planning Regulations between July 1988 and December 1989 identified a number of issues and made various recommen-dations for improvement. Some of the recommendations from 1991 have now been implemented, including the production of a *Good practice guide for the evaluation of environmental information for planning projects* (DoE, 1994e), *Draft guidance on preparing environmental statements for planning projects* (DoE, 1994g), and the requirement on project proponents to submit at least three copies of the ES to the LPA. The 1991 report also included recommendations that research should be undertaken from time to time to ensure that EA procedures were being operated efficiently; this study is one response to this recommendation.

However several other recommendations from the 1991 report have not been followed up to date. The review of the implementation of EC Directive 85/337 by the European Commission (CEC, 1993a) also makes recommendations. The recommendations in the final section of this report are set in the context of these earlier documents, and other suggestions for improving the operation of the EA planning procedures in the UK including the EC Draft Directive on amendments to the 1985 EA Directive (CEC, 1994). They are presented as responses to a number of issues which have emerged from the studies.

7.2.2 Recommendations in response to immediate issues, to improve ES/EA quality

Issue 1: as was discussed in Sections 5.3.1 and 6.3.2, ES quality is related to the extent and nature of pre-submission consultation and scoping activities. Such consultation and scoping was lacking in many cases, and unclear in several others.

- *Recommendation 1.1: DoE Guidance should highlight the importance of the ES reporting on pre-submission consultation and scoping, including with the public (the DoE's Draft Guide (1994g) provides support for this proposal).*

- *Recommendation 1.2: Circular 15/88 should be revised to require the inclusion of a Method Statement in the ES (as recommended in the DoE's Draft Guide (1994g)). This should explain briefly how the study has been conducted in terms of techniques and methods, organisation and timetable, consultation, roles of experts, events involving the general public, and guidance used.*

Issue 2: as was discussed in Section 6.3.3, guidance on EA activity and on the preparation and content of an ES is of considerable value, but has been limited for most ESs produced to date.

- *Recommendation 2.1: the DoE/SoEnD/WO should encourage the dissemination of good practice guidance to the key participants in the EA process.*

- *Recommendation 2.2: LPAs should also encourage the dissemination of good practice. This might be in the form of LPA EA Handbooks. LPAs should also be encouraged to develop planning policies relating to ES quality. Such policies could, for example, be used to give guidance on the issues that would be expected to be addressed in ESs for particular types of development.*

- *Recommendation 2.3: advice on areas of particular weakness in ESs, including the consideration of alternatives and the nature of the non-technical summary, should be emphasized in future DoE Guidance.*

Issue 3: as was discussed in Sections 5.4.2, 5.4.3 and 6.2.2, ES quality is also clearly related to the quality of inputs from environmental consultants and developers, although there is some concern about the objectivity of some inputs.

- *Recommendation 3.1: the certification of competent consultants, and the accreditation of EA training, in discussion with the IEA and the RTPI, should be developed further.*

- *Recommendation 3.2: as part of pre-submission consultation, LPAs and developers should be encouraged to liaise on the most appropriate use of consultants.*

Issue 4: as was discussed in Section 5.3.1, the review of the quality of ESs by LPAs is often very limited, and of little value to other participants in the EA and planning application process.

- *Recommendation 4.1: within time and cost constraints, there may be advantages in the provision by consultants/developers of independent review of their ESs prior to submission to the LPA.*

- *Recommendation 4.2: LPAs should be encouraged to use review frameworks for their assessment of ES quality, as recommended by the DoE (1994g). The findings from such exercises should be available for public scrutiny, including use in decision making in Planning Committee.*

- *Recommendation 4.3: the DoE/SoEnD/WO should amend the Guidance to encourage LPAs and developers to send a full copy of the ES to consultees.*

Issue 5: as was discussed in Sections 4.2 and 5.2, many ESs were poor in terms of effective communication and accessibility to relevant audiences.

- *Recommendation 5.1: the DoE/SoEnD/WO should consider amending the Circular and/or guidance to require that the non-technical summary should be available as a separate document, of not more than 20 pages, for public consumption. This is in addition to its inclusion as the non-technical summary in the full ES.*

- *Recommendation 5.2: the DoE/SoEnD/WO should give consideration to the problem of the high price of many ESs and its implication for public access to information.*

7.2.3 Recommendations in response to associated issues

Associated issue A: as was discussed in section 3.2.5, difficulties were encountered, in many cases, in actually locating, fully identifying and obtaining the ES. The ES document often referred to, and relied on, other material submitted with the application (such as plans).

- *Recommendation A.1: the forthcoming Guidance and the Circular 15/88 should be revised to require a standard front page to be included in each ES, providing at a glance the basic facts about the project, including: submission date, EA regulation(s) and schedule number, name of commissioning organi-sation, consultancies involved in the EA (inc. extent of involvement), local planning authority, location of the project and whether it is submitted as part of an outline or detailed applica-tion. This should be copied and forwarded by the LPA to the DoE/SoEnD/WO.*

- *Recommendation A.2: (as per DoE, 1991) the DoE/SoEnD/WO should amend the Circular 15/88 to make it clear that the ES should be a self-contained document (or documents) presenting all the relevant information relating to the likely impact of a project, and that it should not have to be read in conjunction with the relevant planning application, to obtain certain essential details such as the size and location of the project.*

- *Recommendation A.3: the DoE/SoEnD/WO should investigate further procedures to improve the effectiveness of the ES deposit procedures (in the national repository), develop an ES database, and regularly publish a list of all ESs prepared. The feasibility and utility of establishing a computerised database, possibly in CD-ROM form, for all ESs, merits investigation.*

Associated Issue B: the effects of recent and forthcoming guidance (see Section 6.3.3), and the recommendations noted in 7.2.2, should be assessed.

- *Recommendation B.1: the effects of changes in EA procedures, including guidance, on the quality of ESs and the EA process should be monitored in further research.*

7.2.4 Final Summary

In relation to the history and development of land use planning in the UK the formal requirement for environmental assessment has been with us but a short time. Yet in that time it has become an increasingly important part of the development control process. The importance of EA is reflected in the commitment of the DoE to encourage better practice through the sponsoring of research such as this, and many other EA based studies, as well as the production of practical guidance.

One measure of the development of good practice is the quality of environmental statements and this particular study has demonstrated that the quality of ESs is improving, albeit with some reservations. This improvement is not, perhaps, simply the result of more experience, more guidance and the other factors identified here. Improving quality also reflects the commitment of practitioners to a planning technique that is being recognised increasingly as having benefits for all involved in the development process.

In recognising the improvements in quality it is important not to be complacent and the recommendations set out above are designed to carry this tide of improvements forward. The literature review and particularly the references to procedures and practice in other countries, plus the comments from a number of those surveyed in this study, suggests that there are lessons to be drawn from elsewhere that could be introduced or adapted to help this process further. It should also be noted that at the time of writing the EC are considering amendments to the EA Directive which may include mandatory requirements for early scoping and a consideration of alternatives. These likely changes are of great interest when set against the background of this study for both have been highlighted as being in need of improvement. DGXI are also preparing a Directive requiring environmental assessments of plans, projects and programmes (Strategic Environmental Assessment). The UK has arguably led the way in SEA with the requirement for the environmental appraisal of development plans which was first introduced in 1992 with PPG12 (DoE, 1992c) followed by the 1993 Good Practice Guide (DoE, 1993c). Yet these new Directives will undoubtedly require changes to the UK's EA and planning system.

Ideally the EA process should be an iterative and dynamic process which provides the opportunity of improving projects and safeguarding the environment at every stage. The UK planning system has so far accommodated the introduction of this dynamic process and accepted the changes it has brought. The evidence from those interviewed for this study is that those changes have been for the benefit of the planning process as a whole. As ES quality and EA practice improves and develops so will the benefits to the planning process and the environment increase.

APPENDIX 1: REFERENCES AND BIBLIOGRAPHY

Abracosa, R. and L. Ortolano, (1987) "Environmental Impact Assessment in the Philippines: 1977-1985" *Environmental Impact Assessment Review* 7, pp. 293-310.

Alder, J. (1993) "Environmental Impact Assessment: The inadequacies of English law", *Journal of Environmental Law* 5(2), pp. 203-220.

Bailey, J. and V. Hobbs (1990) "A proposed framework and database for EIA auditing", *Journal of Environmental Management* 31(2), pp. 163-72.

Ball, S. (1991) "Implementation of the environmental assessment Directive in Britain", *Integrated Environmental Management* 5, pp. 9-11.

Ball, S. and S. Bell (1994) *Environmental Law*, Blackstone Press, p. 240.

Bartlit, J.R. (1991) "An adequate EIS under NEPA: deference to CEQ - merely conceptual listing of mitigation leads us to a merely conceptual national environmental policy", *Natural Resources Journal* 31(3), pp. 653-72.

Bedfordshire County Council (1990) "Environmental assessment: A survey for the County Planning Officers Society", Beds CC.

Bisset, R. (1980) "Problems and Issues in the Implementation of EIA Audits", *Environmental Impact Assessment Review* 1(4), pp. 379-396.

Blackmore, M. (1994) *The Effect of Environmental Assessment on Infrastructure Project Planning Decisions*, M. Town Planning dissertation, University of Manchester.

Boon, P. (1990) "Environmental Assessment in the Water Industry", *Proceedings of IWEM Conference*, Cambridge, February.

Brangan, E. (1991) *A Statistical Comparison of EISs Submitted in Ireland between July 1988 and December 1990 with those Submitted in the Same Period in the UK*, Environmental Research Unit, Department of the Environment, Dublin, Ireland.

Braun, C. (1993) "EIA The 1990s: The View From Marsham Street", *EIA in the UK: Evaluation and Prospect Conference*, University of Manchester, 6 April.

Buckley, R.C. (1989) *Precision in Environmental Impact Prediction: First National Environmental Audit, Australia*, Publications Section, Centre for Resource and Environmental Studies, Australian National University, GPO Box 4, Canberra ACT 2601.

Buckley, R.C. (1991) "How accurate are environmental impact predictions?", *Ambio* 20(3-4), pp. 161-2.

Bulleid, P. (1993) "Assessing the Need for EAs", *Planning Week* 17, 18 Nov.

Carnwarth, R. (1991) "The Planning Lawyer and the Environment", *Journal of Environmental Law* 3(1), pp. 57-67.

CBI Planning Task Force (1993) *Shaping the Nation*, CBI, London.

Centre for Environmental Management & Planning (1994) *Public Participation in EIA: A Review of Experience in Europe and the UK*, draft report, CEMP, Aberdeen.

Clark, B. (1992) "A Varied Pattern of Response", *Environmental Assessment and Audit: A Users Guide*, Ambit Publications, Gloucester.

Coles, T. (1993) "Quality in ecological assessments", *Environmental Assessment* 3(1), pp. 71-2.

Coles, T. and J. P. Tarling (1991) *Environmental Assessment: Experience to Date*, Institute of Environmental Assessment, Lincolnshire.

Coles, T. (1990) "Assessing the assessors", *Planning* 892, pp. 26-7.

Coles, T., J. P. Tarling, and K. Fuller (1992) *Practical experience of environmental assessment in the UK*, Institute of Environmental Assessment, Lincolnshire.

Colley, R. and N. Lee (1990) "Reviewing the quality of environmental statements", *The Planner* 76(16), pp.12-4.

Commission of the European Communities (1985) *Directive 85/337/EEC: On the Assessment of the Effects of Certain Public and Private Projects on the Environment*, Official Journal L175, Brussels..

Commission of the European Communities (1991) *A review of the implementation of Directive 85/337/EEC, Volume 1: Main Report*, Contract No. 6610(90)8685.

Commission of European Communities (1992) *Report from the Commission of the Implementation of Directive 85/337/EEC*, Brussels.

Commission of the European Communities (1993a) *Report from the Commission of the Implementation of Directive 85/337/EEC*, COM(93) 28 Final, Vol 13, Brussels, 2 April.

Commission of the European Communities (1993b) *Checklist for the Review of Environmental Information Submitted under EIA Procedures*, DG XI, Brussels.

Commission of the European Communities (1994) *Proposal for A Directive Amending Council Directive 85/337/EEC*, OJ C130, Vol 37, Brussels, 12 May.

Cooper-Kenyon, R. (1991) "Environmental assessment: An overview on behalf of the R.I.C.S.", *Journal of Planning Law*, pp. 419-422.

Council for the Protection of Rural England (1990) *Environmental Statements: getting them right*, London.

Council for the Protection of Rural England (1991) *The Environmental Assessment Directive - Five Years On*, London, May.

Council for the Protection of Rural England (1992) *Mock Directive*, London.

Cowan, A. (1994) "Local authority review and decision making", notes of lecture to MSc/Diploma course in Environmental Assessment, Oxford Brookes University, 22 February.

Culhane, P.J. (1991) "Post-EIS environmental auditing: a first step to making rational environmental assessment a reality", *The Environmental Professional* 15, pp. 66-75.

Dancey, R. and N. Lee (1993) *The Quality of Environmental Impact Statements Submitted in Ireland*, Environmental Research Unit, Dublin.

Dancey, R. and N. Lee (1993) "A comparison of the quality of environmental impact statements in Ireland and the UK", *Environmental Management Ireland* 1(2), pp. 21-6.

Department of the Environment (1988a) *Circular 15/88 (Welsh Office 23/88) Town and County Planning (Assessment of Environmental Effects) Regulations 1988*, HMSO.

Department of the Environment (1988b) *Town and Country Planning (Assessment of Environmental Effects) Regulations (1988)*, HMSO.

Department of the Environment (1988c) *MPG2 Applications permissions and Conditions*, HMSO

Department of the Environment (1989a) *Environmental assessment: a guide to the procedures*. HMSO

Department of the Environment (1989b) *MPG7 The Reclamation of Mineral Workings*, HMSO.

Department of the Environment (1991a) *Monitoring Environmental Assessment and Planning*, HMSO, London.

Department of the Environment (1991b) *MPG10 Provision of Raw Materials for the Cement Industry*, HMSO.

Department of the Environment (1992a) Decision Letter Domestic, Commercial and Industrial Waste: Lewes APP/E1400/A/ 90/ 170677

Department of the Environment (1992b) *PPG1 General Policy and Principles*, HMSO.

Department of the Environment (1992c) *PPG12 Development Plans and Regional Planning Guidance*, HMSO.

Department of the Environment (1993b) *PPG22 Renewable Energy*, HMSO.

Department of the Environment (1993c) *Environmental Appraisal of Development Plans: A Good Practice Guide*, HMSO.

Department of the Environment (1994a) *PPG23 Planning and Pollution Control*, HMSO.

Department of the Environment (1994b) *MPG3 Coal Mining and Colliery Spoil Disposal*, HMSO.

Department of the Environment (1994c) *MPG6 Guidelines for Aggregates Provision in England*, HMSO.

Department of the Environment (1994d) *Good Practice on the Evaluation of Environmental Information for Planning Projects: Research Report,* Land Use Consultants, assisted by the Univ. of East Anglia, HMSO, London.

Department of the Environment (1994e) *Evaluation of Environmental Information for Planning Projects: A Good Practice Guide*, HMSO.

Department of the Environment (1994f) *Town and Country Planning (Assessment of Environmental Effects) (Amendment) Regulations,* SI 1994 No. 677.

Department of the Environment (1994g) *Guide on Preparing Environmental Statements for Planning Projects,* Consultation Draft, DoE, London, July

Department of the Environment/Welsh Office (1989) *Environmental Assessment: A Guide to the Procedures*, HMSO.

Department of Transport (1993) *Design Manual for Roads and Bridges, Vol. 11: Environmental Appraisal*, HMSO.

Dixon, J.E. (1993) "The integration of EIA and planning in New Zealand: Changing process and practice", *Journal of Environmental Planning and Management* 36(2), pp. 239-251.

Dodd, A. (1994) The Environmental Appraisal of Development Plans: the Role and Views of Consultees, in Wilson, E. ed., *Issues in the Environmental Appraisal of Development Plans*, Working Paper No 153, School of Planning, Oxford Brookes University.

Elkin, T.J. and P.G.R. Smith (1988) "What is a good environmental impact statement?", *Journal of Environmental Management* 26, pp. 71-89.

Ensminger, J.T. and R.B. McLean (1993) "Reasons and strategies for more effective NEPA implementation", *The Environmental Professional* 15(1), pp. 46-56.

Environmental Data Services (1993) "Taking Stock of Environmental Assessment", *ENDS Report 221*, June.

Environmental Data Services (1994) "Screening & Scoping Proposed for Environmental Assessment", *ENDS Report* 232, May.

Essex Planning Officers' Association (1994) *Environmental Assessment: The Way Forward. First Draft*, EPOA.

Ferrary, C. (1994) "Environmental Assessment: Our Client's Perspective", *Environmental Assessment: Window Dressing Or?* Conference, Andover, 20 April.

Frost, R. (1994) "Project Design Beyond Environmental Impact Statements", *Environmental Assessment* 2(1) March.

Frost, R. and A. Frankish (1995) *Directory of Environmental Statements: July 1988 - Sept 1994*, Impacts Assessment Unit, School of Planning, Oxford Brookes University.

Fuller, K. (1991) "Reviewing UK's experience in EIA", *Integrated Environmental Management* 1, pp. 12-4.

Fuller, K. (1994) "Reviewing Environmental Statements: Digging Below the Surface", Presentation to *Conference on Assessing Environmental Statements*, Aston University, 18 October.

Ginger, C. and Mohai, P. (1993) "The Role of Data in the EIS Process: Evidence from the BLM Wilderness Review", *Environmental Impact Assessment Review* 13, pp. 109-139.

Glasson, J. and D. Heaney (1993) "Socio-Economic Impacts: the Poor Relations in British Environmental Impact Statements", *Journal of Environmental Planning and Management* 36(3), pp.335-343.

Glasson, J., R. Therivel and A. Chadwick (1994) *Introduction to Environmental Impact Assessment*, UCL Press.

Haeuber, R, (1992) "The World Bank and Environmental Assessment: The Role of Non Governmental Organisations", *Environmental Impact Assessment Review* 12, pp. 331-347.

Her Majesty's Government (1990) *This Common Inheritance*, HMSO.

Haigh, N. (1990) *Manual of Environmental Policy*, Longman.

Hammersley, R. (1994) "Are the Competent Authorities Competent?", paper presented to *Conference on Assessing Environmental Statements*, Aston University, 18 October.

Higham, J.W. and J.C. Day (1990) "The British Columbia offshore exploration environmental assessment: an evaluation", *Impact Assessment Bulletin* 8(1/2), pp. 131-43.

Hinson, P (1994), "Environmental Assessment for windfarms", paper presented to *5th Annual Conference on Advances in EIA*, IBC/EIA, London.

Hopkinson, P.G., J. Bowers and C.A. Nash (1990) *The Treatment of Nature Conservation in the Appraisal of Trunk Roads: Submission to the Standing Advisory Committee on Trunk Road Assessment*, Nature Conservancy Council, Northminster House, Peterborough, PE1 1UA.

Institute of Environmental Assessment/Institute of Terrestrial Ecology (1995a) *Guidelines for Baseline Ecological Assessment*.

Institute of Environmental Assessment/Landscape Institute (1995b) *Landscape and Visual Impact Assessment Guidelines*.

Jain, R.K., L.V. Urban and G.S. Stacey (1981) *Environmental Impact Assessment*, 2nd ed., Van Nostrand Reinhold, New York.

Jones, C. and N. Lee (1993) *Post-Auditing in Environmental Impact Assessment: the Greater Manchester Metrolink Scheme*, Occasional Paper 37, University of Manchester.

Jones, C.E., N. Lee and C. Wood (1991) *UK Environmental Statements 1988-1990: An analysis*, Occasional Paper No. 29, EIA Centre, University of Manchester.

Jones, C. (1994) "EISs - numbers, quality and the EA process", presentation given at *Conference on Assessing Environmental Statements*, Aston University, 18 October.

Jorissen, J. and R. Coenen (1992) "The EEC Directive on EIA and its implementation in the EC member states", in A.G. Colombo, ed., *Environmental Impact Assessment*, Kluwer Academic Publishers.

Kobus, D. and F. Walsh (1994) *The Impact of the Environmental Assessment Process on Mineral Planning Decisions*, Occasional Paper 39, University of Manchester.

La Spina, A. and Sciortino, G. (1993) "Common Agenda, Southern Rules, European Integration and Environmental Change in the Mediterranean States", Liefferink, J.D., Lowe, P.D. and Mol, A.P.J. (eds) *European Integration and Environmental Policy*, Belhaven Press.

Lee, N. (1992) "Improving quality in the assessment process", pp. 12-3 in *Environmental Assessment and Audit: a User's Guide 1992-1993*, Johnston, B. (ed.), Ambit Publications, Gloucester.

Lee, N. (1993) "The quality of environmental statements in the UK", *Integrated Environmental Management* 17, pp. 17-9.

Lee, N. (1994) "Improving performance and strengthening quality control in the EIA process", talk given at *Conference on improving the role of environmental impact assessment in achieving sustainable development*, University of Manchester, 14 April.

Lee, N. and R. Colley (1991) "Reviewing the quality of environmental statements: review methods and findings, *Town Planning Review* 62(2), pp. 239-48.

Lee, N. and C. Jones (1992) Quality control in environmental assessment, *Parliamentary Brief*, June, p. 35.

Lee, N., F. Walsh and G. Reeder (1994) "Assessing the Performance of the EA Process", *Project Appraisal* 9(3), pp. 161-172, Sept.

Lee, N and R. Colley (1990) *Reviewing the Quality of Environmental Statements*, Occasional Paper No. 24, EIA Centre Univ. of Manchester.

Lee, N. and D. Brown (1992) "Quality control in environmental assessment", *Project Appraisal* 7(1), pp. 41-5.

Lee, N. and R. Dancey (1993) "The quality of environmental impact statements in Ireland and the United Kingdom: a comparative analysis", *Project Appraisal* 8(1), pp. 31-6.

Lee-Wright, M. (1994) "Scoping in the Real World", *Planning for the Natural and Built Environment* 1055, 11 Feb.

MacCallum, D.R. (1987) "Follow-up to Environmental Impact Assessment: Learning from the Canadian Government Experience", *Environmental Monitoring and Assessment* 8, pp. 199-215.

McClaren, D. (1993) *The Future for Environmental Impact Assessment: Meeting the Demands of Sustainable Development*, Friends of the Earth.

Melton, J. (1994) *The Role of Statutory Consultation in Environmental Assessment*, MSc in Environmental Assessment and Management dissertation, Oxford Brookes University.

Mertz, S.W. (1989) "The European Economic Community directive on Environmental Assessments: How will it affect the United Kingdom Developer?", *Journal of Planning Law*, pp. 483-498.

Miller, S. (1993) "Environmental impact statements: whither they go?", *Environmental Science and Technology* 27(7), pp. 1248-9.

Mills, J. (1994) "The adequacy of visual impact assessments in environmental impact statements", in R. Therivel (ed.) *Issues in Environmental Impact Assessment*, Working Paper No. 144, School of Planning, Oxford Brookes University.

Ministry for the Environment, New Zealand (1991) *Guide for Scoping and Public review Methods in Environmental Impact Assessment*, MFE, Wellington.

Nelson, P. (1994) "Guidance for better environmental impact assessment", talk given at *Conference on improving the role of environmental impact assessment in achieving sustainable development*, University of Manchester, 14 April.

Nelson, P. (1995) "Better guidance for better EIA", *Built Environment Journal*, Vol 20, No. 4.

Nicholson, K. (1993) "On the receiving end: dealing with environmental statements", *Environmental Assessment* 1(2), pp. 38-9.

Nordhav, Y. (1992) "Americans Look Askance at Assessment Practices", *Planning* 996, 27 Nov.

O'Sullivan, M. (1991) *Environmental Impact Assessment: A Handbook*, Resource and Environmental Management Unit, University College Cork, Ireland.

Organisation for Economic Co-operation and Development (1991) *Good Practices for Environmental Impact Assessment of Development Projects*, OECD/GD (91)200, Development Assistance Committee, OECD, Paris.

Ortolano, L. (1993) "Controls on project proponents and environmental impact assessment effectiveness", *The Environmental Professional* 15, pp. 352-363.

Petts, J. and G. Eduljee (1994) *Environmental impact assessment for waste treatment and disposal projects*, Wiley.

Planning (1995) 'Consultancy market is squeezed in the middle', No.1117, May.

R v Poole Borough Council ex parte Beebe and others (1991) *Journal of Planning and Environmental Law*, p.643.

Rawlinson, C. (1990) "The impact of environmental assessment", *Estates Gazette* 9014, pp. 20-1.

Rivas, V., A. Gonzalez, D.W. Fischer and A. Cendrero (1994) "An approach to environmental assessment within the land-use planning process: Northern Spanish experiences", *Journal of Environmental Planning and Management* 37(3), pp. 305-322.

Ross, W.A. (1987) "Evaluating Environmental Impact Statements", *Journal of Environmental Management*, 25, pp. 137-147.

Sadler, B. (1994) *International study of the effectiveness of environmental assessment: proposed framework*, FEARO, Canada.

Salter, J.R. (1992) "Environmental Assessment: The challenge from Brussels", *Journal of Planning Law*, pp.14 20.

Schibuola, S. and P.H. Byer (1991) "Use of knowledge-based systems for the review of environmental impact statements", *Environmental Impact Assessment Review* 11(1), pp. 11-28.

Sheate, W. (1992) "Lobbying for effective environmental assessment", *Long Range Planning* 25(4), pp. 90-8.

Sheate, W. (1994) *Making an Impact: A Guide to EIA Law and Policy*, Cameron May, London.

Simpson, B. (1994) "Chairman's Introduction", *Assessing Environmental Statements Conference*, Aston University, 18 Oct.

Skehan, D.C. (1993) "EIA in Ireland", presentation given to conference on *Advances in Environmental Impact Assessment*, organised by Institute of Environmental Assessment and IBC Technical Services Ltd., 25-26 November.

Southerland, M.T. (1992) "Consideration of terrestrial environments in the review of environmental impact statements", *The Environmental Professional* 14(1), pp. 1-9.

Street, E. (1993) "Notes from the coal-face on environmental assessment", *Planning* 1022, 11 June.

Tarling, J. (1991) *A comparison of Environmental Assessment procedures and experience in the UK and the Netherlands*, dissertation, MSc in Environmental Management, University of Stirling.

Therivel, R. and D. Heaney (1994) "Does the sum of project EIAs equal SEA? A transport example", unpublished research report, Oxford Brookes University.

Therivel, R., and J. Weston (1994) *EIA Update for Ambit Publications*, Oxford, March.

Therivel, R., E. Wilson, S. Thompson, D. Heaney and D. Pritchard (1992) *Strategic Environmental Assessment*, Earthscan.

Tomlinson, P. (1989) "Environmental statements: Guidance for review and audit", *The Planner*, 3 November, pp. 12-15.

Treweek, J., S. Thompson, N. Veitch and C. Japp (1993) "Ecological assessment of proposed road developments: a review of environmental statements", *Journal of Environmental Planning and Management* 36(3), pp. 295-307.

Tromans, S. (1991) "Town and Country Planning and Environmental Protection". *Journal of Planning and Environmental Law*, pp. 6-42.

US Environmental Protection Agency (1993) *Habitat Evaluation: Guidance for the Review of Environmental Impact Assessment Documents*, EPA, Washington DC, 20460.

Vale of White Horse District Council (1993) *Local Plan: Draft for Consultation*, VWHDC.

Wathern, P. (1988) *Environmental impact assessment: theory and practice*, Unwin Hyman.

West, C., R. Bissett, and R. Snowden (1993) "Developing Countries EIAs", presentation given to *Conference on Advances in Environmental Impact Assessment*, organised by Institute of Environmental Assessment and IBC Technical Services Ltd., 25-26 November

Weston, J. (1994) "Assessments at the appeal cutting edge", *Planning* 1075, p. 24.

Weston, J. (1996 forthcoming) Consultants in the EIA process'.

Williams, R. H. (1988) "The Environmental Impact Assessment Directive of the European Communities" in Clark, M. and Herington, J., eds., *The Role of Environmental Impact Assessment in the Planning Process*, Mansell Publishing Ltd.

Winter, P. (1994) "Planning and sustainability: an examination of the role of the planning system as an instrument for the delivery of sustainable development", *Journal of Planning Law*, pp. 883-900.

Wood, C. (1991) "Environmental Impact Assessment in the United Kingdom", presentation given at the *ACSP-AESOP Joint International Planning Congress*, Oxford, July.

Wood, C. (1993a) "EIA in Europe - Results of the Five Year Review" presentation given to conference on Advances in Environmental Impact Assessment, organised by Institute of Environmental Assessment and IBC Technical Services Ltd., 25-26 November

Wood, C. (1993b) "Antipodean Environmental Assessment", *Town Planning Review* 64(2), pp. 119-138.

Wood, C. (1994a) "Evaluating impact assessment systems", *Integrated Environmental Management* 30, pp. 10-3.

Wood, C. (1994b) "Five years of British environmental assessment: an evaluation", Part VI in D. Cross and C. Whitehead, eds., *Development and Planning 1994*, Department of Land Economy, University of Cambridge.

Wood, C. and Bailey, J. (1994) "Predominance and Independence in Environmental Impact Assessment: The Western Australian Model", *Environmental Assessment Review* 14, pp. 37-59.

Wood, C. and C. Jones (1992) "The Impact of Environmental Assessment on local planning authorities", *Journal of Environmental Planning and Management* 35(2), pp. 115-127.

Wood, C. and C. Jones (1994) "Increasing the Effect of EIA on Planning Decisions", presentation to *Conference on The Role of Environmental Impact Assessment in Achieving Sustainable Development*, EIA Centre, University of Manchester, 14 April.

Wood, C. and C. Jones (1995 unpublished) "The Effect of Environmental Assessment on Planning Decisions". Seminar presentation, Univ. of Manchester (July 95).

Wood, C., N. Lee and C. Jones (1991) "Environmental statements in the UK: the initial experience", *Project Appraisal* 6(4), pp. 187-95.

World Bank, (1991) *Operational Directive 4.01: Environmental Assessment*, World Bank.

APPENDIX 2: INITIAL QUESTIONNAIRE SENT TO EXPERIENCED EA PRACTITIONERS

Respondents

National Power

Agricultural Development Advisory Service

National Rivers Authority

Institute of Environmental Assessment

Environmental Resource Management

Arup Environmental

David Tyldesley & Assoc

Nicholas Pearson & Assoc

RPS Clouston

Hampshire CC

Clwyd CC

Cheshire CC

Cambridgeshire CC

Impacts Assessment Unit   School of Planning

OXFORD
BROOKES
UNIVERSITY

JG/mp

23 January 1995

Gipsy Lane Campus
Headington Oxford OX3 0BP

Tel: 01865 483450
Fax: 01865 483559
Direct Line: 01865 483

Head: Professor John Glasson
BSc(Econ) MA MRTPI MIMgt FRSA

Dear

DOE Research Project: Changes in the Quality of Environmental Statements for Planning Projects

The Impacts Assessment Unit is undertaking the above research to show <u>what changes</u>, if any, there have been in the quality of ESs for planning projects in England, Scotland and Wales since the earlier Manchester University study (published by the DOE 1991), and to explain <u>why such changes</u>, if, any have taken place.

We are in the initial 'literature review' stage of the project. We are familiar with much of the published literature of relevance in the UK, and elsewhere. However we should like to cover also any semi-published or in house literature on the changing quality of ESs from some of the most experienced EA practitioners in consultancies, industry and local government in the UK.

I realise that you do receive many requests for information on your EA activities. The detailed request for information, on the reverse of this letter, is very brief and I would be grateful if you could spend just a few minutes to run through the questions. I enclose a stamp addressed envelope for your reply.

Yours sincerely

Prof John Glasson
Head of School of Planning

Contact: _____

Respondent (if not contact above): _____

DOE Research Project: Changes in the Quality of Environmental Statements for Planning Projects in England, Scotland and Wales

1. We are particularly interested in the issue of 'quality for whom', involving a consideration of the ES from the perspectives of the various parties in the EA process. What criteria do <u>you</u> employ to assess the quality of the assessment as reported in an ES? (Max 5).

   _____
   _____
   _____
   _____
   _____

2. When preparing/reviewing an ES do you make use of any 'best practice guidance', and if yes, which?

   _____
   _____
   _____
   _____
   _____

3. Do you have any in-house quality control systems for ESs which you produce/review? If yes, would it be possible for us to: [Yes]   [No]

   [a] have a copy of your documentation?   [Yes]   [No]

   [b] discuss your system(s) with you by phone?   [Yes]   [No]

4. Has _____ undertaken any review of the changing quality of:

   [a] • ESs produced/reviewed by your consultantcy
        • ESs produced by/for..............................
        • ESs received by ....   [Yes]   [No]

   [b] the wider population of ESs produced in the UK since the introduction of EA regulations in 1988?   [Yes]   [No]

5. If 'yes' to 4a and/or 4b, would it be possible for us to:

   [a] have a copy of the findings?   [Yes]   [No]

   [b] discuss the findings with you by phone?   [Yes]   [No]

   If 'Yes' to 3a and/or 5a, could you please forward any documentation to us in the envelope provided, noting any confidentiality restrictions you might wish to apply.

6. What do <u>you</u> think are the <u>main determinants</u>, which might explain any changes in the quality of ESs in the UK since 1988? (Max.5):

   _____
   _____
   _____
   _____

Thank you for your assistance. Can you please return this form with any relevant documentation, in the enclosed envelope, to the Impacts Assessment Unit, School of Planning, Oxford Brookes University, Headington, Oxford, OX3 0BP

APPENDIX 3: OTHER ES REVIEW FRAMEWORKS

3.1 Simple 'regulatory requirements' criteria

3.2 Lee and Colley

3.3 EU Review Checklist (EC DGXI)

3.1 Simple 'regulatory requirements' criteria (based on Para 2 of Schedule 3 of the TCP(AEE) 1988 Regulations)

1. Describes the proposed development, including its design and size or scale.

2. Defines the land area taken up by the development site and any associated works, and shows their location on a map

3. Describes the uses to which this land will be put, and demarcates the land use areas.

4. Considers direct and indirect effects of the project, and any consequential development.

5. Investigates these impacts in so far as they affect human beings, flora, fauna, soil, water, air, climate, landscape, interactions between the above, material assets, and cultural heritage.

6. Considers the mitigation of all significant negative impacts.

7. Mitigation measures considered include modification of the project, the replacement of facilities, and the creation of new resources.

8. There is a non-technical summary which contains at least a brief description of the project and environment, the main mitigation measures, and a description of any remaining impacts.

9. The summary presents the main findings of the assessment and covers all the main issues raised.

3.2 Lee and Colley review package

The Lee Colley Method reviews ESs under four main topics, each of which is examined under a number of sub-headings:

(i) **Description of the development, the local environment, and the baseline conditions:**
   - Description of the development
   - Site description
   - Residuals
   - Baseline conditions

(ii) **Identification and evaluation of key impacts:**
   - Identification of impacts
   - Prediction of impact magnitudes
   - Assessment of impact significance

(iii) **Alternatives and mitigation:**
   - Alternatives
   - Mitigation
   - Commitment to mitigation

(iv) **Communication of results:**
   - Presentation
   - Balance
   - Non-technical summary

In outline, the content and quality of the environmental Statement is reviewed under each of the subheads, using a sliding scale of Assessment Symbols A-F:

Grade A indicates that the work has generally been well performed with no important omissions;
Grade B is generally satisfactory and complete with only minor omissions and inadequacies;
Grade C is regarded as just satisfactory despite some omissions or inadequacies;
Grade D indicates that parts are well attempted but, on the whole, just unsatisfactory because of omissions or inadequacies;
Grade E is not satisfactory, revealing significant omissions or inadequacies;
Grade F is very unsatisfactory with important task(s) poorly done or not attempted.

Having analysed each subhead, aggregated scores are given to the four review areas, and a final summary grade is attached to the whole statement.

3.3 EU Review Checklist (EC DGXI)

1. **Objectives of the Review Checklist**

    This Review Checklist has been developed as a method for reviewing environmental information submitted by developers, to the competent authorities, as part of an EIA procedure[1]. Its purpose is to assist reviewers in evaluating the completeness and suitability of this information from a technical and decision making viewpoint. In particular it will assist reviewers in deciding whether all relevant information is available to fulfil two main functions:

    - to provide an adequate basis for decision making;
    - to provide information to the public.

    The checklist is not intended to be a tool to verify whether the information provided meets legal requirements. This is only possible in the context of specific Member States' legislation.

    The use of the Review Checklist is entirely voluntary. It is not meant to replace any existing review method in the Member States. Reviewers may use it where they consider it helpful. When using it reviewers are free to adapt it to suit local circumstances or to reflect practical experience over time. Some provisions for this have been made in the Checklist.

2. **Review Criteria**

    The review criteria are organised in eight *Review Areas* as follows:
    - 1 Description of the Project
    - 2 Outline of Alternatives
    - 3 Description of the Environment
    - 4 Description of Mitigation Measures
    - 5 Description of Effects
    - 6 Non-Technical Summary
    - 7 Difficulties Compiling Information
    - 8 General Approach

    Within each Review Area there are *Review Questions* which identify, in some detail, the items of information which may need to be provided by the developer to the competent authority.

3. **Review Method**

    In general there are four sequential tasks to be carried out.

    - Firstly, decide which information is relevant in the specific context of the project.

    - Secondly, determine whether there are omissions and/or shortcomings in the information presented.

    - Thirdly, if there are omissions and/or shortcomings, determine which of these are crucial for the decision making process.

    - Fourthly, specify which additional information is required and, where appropriate and feasible, recommend a way of obtaining this information.

The checklist has been designed on the basis of these tasks. The Review method consists of the following steps:

Step 1 (the first task):

For each Review Question the reviewer first of all decides whether the particular type of information is relevant to the type of development proposed. If not, the reviewer notes this and moves on to the next Question.

Step 2 (the second and third tasks):

If the Question is considered relevant the reviewer examines the information provided by the developer and assesses it as:

- **Complete:** all information relevant to the decision-making process is available; no additional information is required;

- **Acceptable:** the information presented is not complete, however, the omissions need not prevent the decision-making process proceeding;

- **Inadequate:** the information presented contains major omissions; additional information is necessary before the decision-making process can proceed.

Guidance on factors to be considered in assessing the adequacy of information is given in the "Guidance Notes" section.

Step 4 (the fourth task):

Where a Question is assessed as Acceptable or Inadequate, the reviewer notes in the right hand column what information is missing, and, where appropriate and feasible, recommends a way of obtaining this information.

[1] In some jurisdictions this information may be presented in an Environmental Impact Statement comprising one or more documents. In others the information may be provided in other documents and formats, for example as part of an application for development consent.

APPENDIX 4: LIST OF 25 MATCHED PAIRS - CHARACTERISTICS

No.	Type of Project	Location/Description	Date	Schedule	Oxf.	Dept.	County
1a	Road	A130 Bypass (A12 - A132)	5.90	2.10d	yes		Essex
1b	Road	Colchester Eastern Approaches Stage 2	22.7.93	2.10d	yes		Essex
2a	Road	Woodstock bypass	5.12.90	2.10d	yes		Oxfordshire
2b	Road	B4031 Finmere Diversion	4.94	2.10c	yes		Oxfordshire
3a	Waste disposal - landfill	Reilly Quarry, Bishopton	27.10.81	2.11c	yes	SO	Strathclyde
3b	Waste disposal - landfill	Bedlay Colliery, Mollinsburn	7.2.94	2.11c	yes	SO	Stralhclyde
4a	Waste disposal - landfill extension	Ufton Hill Farm	18.1.90	2.11c	yes	DoE	Warwickshire
4b	Waste disposal - landfill extension	Burton Farrn, Bishopton	4.92	2.11c	yes	DoE	Warwickshire
5a	Waste disposal - landfill	Northfleet	.90	2.11c	yes		Kent
5b	Waste disposal - landraise	North Farm, Tunbridge Wells	5.92	2.11c	yes		Kent
6a	Sewage treatment works	Teign Estuary	12.90	2.11d	yes	DoE	Devon
6b	Sewage treatmnent works	Taw Torridge, Northern Area	7.93	2.11d	yes	DoE	Devon
7a	Waste disposal - landfill	Roxby	8.89	1.92(2)	yes	DoE	Humberside
7b	Waste disposal - landfill (special waste)	Courtaulds factory, Grimsby	10.6.93	1.9(2)	yes		Humberside
8a	Waste treatment works	Pitsea	.88	1.9	yes		Essex
8b	Waste recycling	Poynters Lane, N. Shoeburyness	1.92	2.11c	yes	DoE	Essex
9a	Extraction - sand & gravel	Passenham, Stony Stratford	88/91	2.2c	yes		Northants
9b	Extraction - sand & gravel	Irthlingborough	16.1.92	2.2c	yes		Northants
10a	Extraction - sand & gravel	Spring Farrn	.90	2.2c	yes		N. Yorkshire
10b	Extraction - sand & gravel	Manor House Farm, Ellerton-on-Swale.	7.92	2.2c	yes		N. Yorkshire
11a	Opencast coal	Helid Colliery	10.89	2.2d	yes		Mid Glamorgan
11b	Opencast coal	Selar	25.6.93	2.2d	yes		Mid/West Glam
12a	Opencast coal	Airdsgreen	14.2.89	2.2d	yes	SO	Strathclyde
12b	Opencast coal - extension	Kittymuirhill Farm	19.2.92	2.2d	yes	SO	Strathclyde
13a	Industrial - timber	Morayhill Inverness	22.5.90	2.8c	yes	SO	Highland
13b	Industrial - paper	Morayhill	17.6.92	2.8c	yes	SO	Highland
14a	New settlement	Highfields	5.89	2.10b	yes	DoE	Cambridgeshire
14b	New settlement	Monkfield Park	9.92	2.10b	yes		Cambridgeshire
15a	Industrial - light	Belmont Industrial Estate	19.12.90	2.10a	yes		Durham
15b	Industrial Estate	Dawdon Colliery	23.8.93	2.10b	yes	DoE	Durham
16a	Mixed use development	Priory Park	8.1.90	2.10b	yes		Greater London
16b	Mixed use development	Kings Cross	5.92	2.10b	yes		Greater London
17a	Mixed use development	Greatwood, Llanwern	17.12.90	2.10b	yes		Gwent
17b	Mixed use development	Gwent Euro Park	26.8.92	2.10b	yes		Gwent
18a	Power station - tyres	Four Ashes, Cannock	11.90	2.3a	yes	DoE	Staffordshire
18b	Power station - CHP, tyre recycling	nr. Leicester	12.8.93	2.3a	yes		Leicestershire
19a	Windfarm	Deli Farm, Delabole	8.89	2.3k	yes	DoE	Cornwall
19b	Windfarm	Worthyvale Manor / Waterpit Down.	12.93	2.3k	yes		Cornwall
20a	Windfarm	Kirkby Moor	12.9.91	2.3k	yes		Cumbria
20b	Windfarm	Silloth	24.11.93	2.3k			Cumbria
21a	Leisure	Fullwood Farm	19.4.90	2.11a	yes	DoE	Derbyshire
21b	Leisure	Environmental Leisure Centre	3.92.	2.11a	yes		Derbyshire
22a	Agriculture - turkeys	Bar Farm, Deeping Fen	24.4.89	2.1b	yes		Lincolnshire
22b	Agriculture - poultry	Frieston, Boston	27.1.93	2.1b	yes		Lincolnshire
23a	Food processing	Sheet Road, Ludlow	7.89	2.7	yes	DoE	Shropshire
23b	Food - extension	Project Alaska, Allscott, Telford	4.3.94	2.7	yes	DoE	Shropshire
24a	Reservoir	Blashford Lakes Stage II	2.91	2.10f	yes		Hampshire
24b	Reservoir	Longham Waterworks	10.92	2.10f			Dorset
25a	Motorway service area	Strensham-upon-Avon	23.5.89	2.10b,2.10	yes	DoE	Heref. & Worcs.
25b	Motorway service area	Warden Heath Farm	.92	2.10k			Staffordshire

No.	Local Authority	Developer/Agent	Appeal	Approved
1a	Essex CC	Essex CC, Highways Dept.		Yes - 1.90
1b	Essex CC	Essex CC, Highways Dept.		Yes - 22.10.93
2a	Oxfordshire County Council	Oxon CC, Highways Dept		Yes - 6.8.91
2b	Oxfordshire CC	Oxfordshire CC		Yes
3a	Renfrew DC	Tarmac Econowaste Ltd		Yes - 2.7.92
3b	Strathkelvin DC (Monklands DC consultee)	Alexander Russel Plc		Yes - 30.9.94
4a	Warwickshire County Council	Mr P Mann, c/o M J Carter Assoc		Yes
4b	Warwickshire County Council	Burton Farms (Stratford) Ltd co Applied Geology (central) Ltd		Refused
5a	Kent CC	Blue Circle / Kent CC, Waste Disposal Div.		Yes
5b	Kent CC	Kent CC		Yes - 20.1.93
6a	Teignbridge DC	South West Water Services Ltd, c/o ERL		Yes - 13.5.91
6b	Devon CC	South West Water Services Ltd. co WS Atkins		Yes-9.9.94
7a	Humberside CC	Waste Management Ltd		Yes
7b	Humberside CC	Courtaulds		Yes
8a	Essex CC	Cleanaway		Yes
8b	Essex CC	Cityscape Ltd/Adkins, Goreham and Richardson		Refused - 26.10.92
9a	Northamptonshire County Council	Hall Aggregates (Eastern Counties) Ltd		Refused - 26.10.90
9b	Northamptonshire CC	ARC Central	Yes	Decision pending
10a	North Yorkshire CC	Rombus Materials Ltd.		Refused
10b	North Yorkshire CC	Ennemix Construction Materials Ltd co Geoplan Ltd	Yes	
11a	Mid Glam. CC (Rhymney Valley DC consultee)	British Coal opencast - South Wales Region		Yes - 8.91
11b	Mid Glamorgan CC, West Glamorgan CC	British Coal Opencast (South Wales Region)		Yes
12a	Cumnock and Doon Valley DC	British Coal Opencast/co Technology Ltd		Yes - 2.12.89
12b	Hamilton DC	Law Mining Ltd		Yes - 22.7.92
13a	Highland RC	Highland Forrest Products Ltd co Environmental Dynamics		Yes - 31.10.90
13b	Highland RC	Norbord		Yes - 9.12.92
14a	South Cambridgeshire District Council	c/o Barton Wilmore	Yes	Refused - 5.3.92
14b	South Cambridgeshire District Council	Alfred McAlpine Projects Ltd		Yes
15a	Durham CC	Durham CC		Yes - 3.6.91
15b		Durham CC		Yes - 31.1.94
16a	London Borough of Merton	G L Hearn & Partners		Yes - 1.11.90
16b	London Borough of Camden	Railways Lands Group		Withdrawn - 6.92
17a	Newport BC			Refused 19.12.91
17b	Monmouth Borough Council	Gwent Euro Park & Tesco		Yes - 6.1.93
18a	South Staffordshire DC	Adams Integrated Waste Ltd/ C L Associated		Yes - 26.2.91
18b	Leicester City Council	Leicester Combined Heat & Power Ltd co Rendel Planning		Refused
19a	North Cornwall DC	Deli Farm	Inquiry	Yes - 23.3.90
19b	North Cornwall DC	Worthyvale Manor Partnership		Refused
20a	South Lakeland DC	Wind Energy Group		Yes - 11.3.92
20b	Allerdale DC	International Wind Development UK	Called in	Withdrawn
21a	Derbyshire Dales DC	Eimport Development Ltd co Chris Cowen Partnership		Yes - 14.5.90
21b	Chesterfield BC	Ecodome UK Ltd		Yes - 14.9.92
22a	South Holland DC	Poddingtons co Kelsey Assoc.		Yes - 22.8.91
22b	Boston BC	Turners Turkeys		Yes - 21.12.93
23a	South Shropshire District Council	Margetts Food Ltd		Yes - 19.12.89
23b	Wrekin District Council	British Sugar plc. co HMY Landscape		Yes
24a	Hampshire CC	Wessex Water Plc co Nicholas Pearson Assoc.		Yes - 30.12.91
24b	Dorset CC	Bournemouth and West Hampshire Water Co Ltd		Yes - 19.4 .94
25a	Wychavon BC	The Lands Improvement Group Ltd, c/o Bodfan Gruffydd Partner.		Refused - 23.11.89
25b	Stafford Borough Council	Granade Hospitality/Terence O'Rourke		Yes

APPENDIX 5: REVIEW FRAMEWORK FOR THE STUDY

EIS number:

project name:

reviewer name:

marking criteria
x√ column: x or √ whether minimum legal criteria have Seen fulfilled (only where min column has a *)
IAU/LC column: use symbols from last page (A-F) to summarise how well EIS fulfils criterion for <u>all</u> criteria
EU column use EU criteria (c,a,i,nr)

1. DESCRIPTION OF THE DEVELOPMENT

min	EU / IAU	LC	criterion	performance against criteria			comments
				√ x	IAU/ LC	EU	
	Principal features of the project						
	1.1	1.1.1	'Explains the purpose(s) and objectives of the development.				
	1.2		Indicates the nature and status of the decision(s) for which the environmental information has been prepared				
	1.3	1.2.3	Gives tbe estimated duration of the construction, operational and, wHere appropriate, decommissioning phase, and the programme within tbese phases.				
*	1.4	1.1.2	Describes the proposed development, including its design and size or scale. Diagrams, plans or maps will usually be necessary for this purpose.				
		1.1.3	Indicates the physical presence or appearance of the completed development within the receiving environment.				
	1.5		Describes the methods of construction.				
	1.6		Describes the nature and methods of production or other types of activity involved in the operation of the project.				
	1.7		Describes any additional services (water, electricity, emergency services etc.) and developments required as a consequence of the project.				
			Describes the project's potential for accidents, hazards and emergencies.				
	Land requirements						
*	1.8, 1.9	1.2.1	Defines the land area taken up by the development site and any associated arrangements, auxiliary facilities and landscaping areas and by the construction site(s), and shows their location clearly on a map. For a linear project, describes the land corridor, vertical and horizontal alignment and need for tunnelling and earthworks.				
*	1.10	1.2.2	Describes the uses to which this land will be put, and demarcates the different land use areas.				
	1.11		Describes the reinstatement and after-use of landtake during construction.				

min	EU	LC	criterion	performance against criteria			comments
	IAU			x √	IAU/LC	EU	
Project Inputs							
	1.16	1.1.4	Describes the nature and quantities of materials needed during the construction and operational phases.				
	1.17	1.2.4a	Estimates the number of workers and visitors entering the project site during both construction and operation.				
	1.18	1.2.4b	Describes their access to the site and likely means of transport				
	1.19		Indicates the means of transporting materials and products to and from the site during construction and operation, and the number of movements involved.				
Residues and emissions							
	1.12	13.1	Estimates the types and quantities of waste matter, energy (noise, vibration, light, heat, radiation etc.) and residual materials generated during construction and operation of the project, and rate at which these will be produced.				
	1.13	1.32	Indicates how these wastes and residual materials are expected to be handled/treated prior to release/disposal, and the routes by which they will eventually be disposed of to the environment.				
	1.14		Identifies any special or hazardous wastes (defined as..) which will be produced, and describes the methods for their disposal as regards their likely main evironmental impacts.				
	1.15	1.3.3	Indicates the methods by which the quantities of residuals and wastes were estimated. Acknowledges any uncertainty, and gives ranges or confidence limits where appropriate.				

overall mark:

2. DESCRIPTION OF THE ENVIRONMENT

min	EU	LC	criterion	performance against criteria			comments
	IAU			√ x	IAU/ LC	EU	
	Description of the area occupied by and surrounding the project						
	3.1	1.4.1	Indicates the area expected to be significantly affected by the various aspects of the project with the aid of suitable maps. Explains the time over which these impacts are likely to occur.				
	3.2		Describes the land uses on the site(s) and in surrounding areas.				
	3.3	1.4.2	Defines the affected environment broadly enough to include any potentially significant effects occurring away from the immediate areas of construction and operation. These may be caused by, for example, the dispersion of pollutants, infrastructural requirements of the project, traffic etc.				
	Baseline conditions						
	3.4	1.5.1a	Identifies and describes the components of the affected environment potentially affected by the project				
	3.5	1.5.1b	The methods used to investigate the affected environment are appropriate to the size and complexity of the assessment task. Uncertainty is indicated.				
	3.6		Predicts the likely future environmental conditions in the absence of the project. Identifies variability in natural systems and human use.				
	3.7	1.5.2	Uses existing technical data sources, including records and studies carried out for environmental agencies and for special interest groups.				
	3.8	1.5.3	Reviews local, regional and national plans and policies, and other data collected as necessary to predict future environmental conditions. Where the proposal does not conform to these plans and policies, the departure is justified.				
	3.9		Local, regional and national agencies holding information on baseline environmental conditions have been approached.				

Overall mark:

3. SCOPING, CONSULTATION, AND IMPACT IDENTIFICATION

min	EU IAU	LC	criterion	performance against criteria √ x		IAU/ LC	EU	comments
Scoping and consultation								
		2.3.1, 2.3.2	There has been a genuine attempt to contact the general public, relevant public agencies, relevant experts and special interest groups to appraise them of the project and its implication. Lists the groups approached.					
			Statutory consultees have been contacted. Lists the consultees approached.					
			Identifies valued environmental attributes on the basis of this consultation.					
		2.3.3	Identifies all project activities with significant impacts on valued environmental attributes. Identifies and selects key impacts for more intense investigation. Describes and justifies the scoping methods used.					
			Includes a copy or summary of the main comments from consultees and the public, and measures taken to respond to these comments.					
Impact identification								
*	5.1. 5.2	2.1.1a	Considers direct and indirect/secondary effects of constructing, operating and, where relevant, after-use or decommissioning of the project (including positive and negative effects). Considers whether effects will arise as a result of 'consequential' development.					
*	5.3 (a)	2.1.2	Investigates the above types of impacts in so far as they affect: human beings, flora, fauna, soil, water, air, climate, landscape, interactions between the above, material assets, cultural heritage.					
	5.3 (b)		Also noise, land use, historic heritage, communities.					
	5.4		If any of the above are not of concern in relation to the specific project and its location, this is clearly stated.					
	5.7	2.2.1, 2.2.2	Identifies impacts using a systematic methodology such as project specific checklists, matrices, panels of experts, extensive consultations, etc. Describes the methods/approaches used and the rationale for using them.					
	5.5		The investigation of each type of impact is appropriate to its importance for the decision, avoiding unnecessary information and concentrating on the key issues.					
	5.6		Considers impacts which may not themselves be significant but which may contribute incrementally to a significant effect.					
	5.8	2.1.3	Considers impacts which might arise from non-standard operating conditions, accidents and emergencies.					
	5.9		If the nature of the project is such that accidents are possible which might cause severe damage within the surrounding environment, an assessment of the probability and likely consequences of such events is carried out and the main findings reported.					

Overall mark:

4. PREDICTION AND EVALUATION OF IMPACTS

min	EU	LC	criterion	performance against criteria			comments
	IAU			√ x	IAU/ LC	EU	
Prediction of magnitude of impacts							
	5.10		Describes impacts in terms of the nature and magnitude of the change occurring and the nature location, number, value, sensitivity) of the affected receptors.				
	5.11	2.1.1b	Predicts the timescale over which the effects will occur, so that it is clear whether impacts are short, medium or long term, temporary or permanent, reversible or irreversible.				
	5.12	2.4.3	Where possible, expresses impact predictions in quantitative terms. Qualitative descriptions, where necessary, are as fully defined as possible.				
		5.13	Describes the likelihood of impacts occurring, and the level of uncertainty attached to the results.				
Methods and data							
	5.14	2.4.2	The methods used to predict the nature, size and scale of impacts are described, and are appropriate to the size and importance of the projected disturbance.				
	5.18	2.4.1	The data used to estimate the size and scale of the main impacts are sufficient for the task, clearly described, and their sources clearly identified. Any gaps in the data are indicated and accounted for.				
Evaluation of impact significance							
	5.16		Discusses the significance of effects in terms of the impact on the local community (including distribution of impacts) and on the protection of environmental resources.				
	5.18		2.5.3 Discusses the available standards, assumptions and value systems which can be used to assess significance.				
	5.19		Where there are no generally accepted standards or criteria for the evaluation of significance, alternative approaches are discussed and, if so, a clear distinction is made between fact, assumption and professional judgement.				
	5.17	2.5.2	Discusses the significance of effects taking into account the appropriate national and international standards or norms, where these are available. Otherwise the magnitude, location and duration of the effects are discussed in conjunction with the value, sensitivity and rarity of the resource.				
		2.1.4	Differentiates project-generated impacts from other changes resulting from non-project activities and variables.				
	5.20		Includes a clear indication of which impacts may be significant and which may not.				

Overall mark:

5. ALTERNATIVES

min	EU	LC	criterion	performance against criteria			comments
		IAU		√ x	IAU/ LC	EU	
	2.1	3.1.1a, 3.1.2	Considers the "no action" alternative, alternative processes, scales, layouts, designs and operating conditions where available at an early stage of project planning, and investigates their main environmental advantages and disadvantages.				
		3.1.3	If unexpectedly severe adverse impacts are identified during the course of the investigation, which are difficult to mitigate, alternatives rejected in the earlier planning phases are re-appraised.				
	2.2	3.1.1b	Gives the reasons for selecting the proposed project, and the part environmental factors played in the selection.				
	2.3		The alternatives are realistic and genuine.				
	2.4		Compares the alternatives' main environmental impacts clearly and objectively with those of the proposed project and with the likely future environmental conditions without the project.				

Overall mark:

6. MITIGATION AND MONITORING

min	EU	LC	criterion	performance against criteria			comments
	IAU			√ x	IAU/LC	EU	
Description of mitigation measure							
*	4.1 (a)	3.2.1a	Considers the mitigation of all significant negative impacts and, where feasible, proposes specific mitigation measures to address each impact.				
*	4.4	3.2.2	Mitigation measures considered include modification of project design, construction and operation the replacement of facilities/resources, and the creation of new resources, as well as 'end-of-pipe' technologies for pollution control.				
	4.2		Describes the reasons for choosing the particular type of mitigation, and the other options available.				
	4.5, 4.6	3.2.3	Explains the extent to which the mitigation methods will be effective. Where the effectiveness is uncertain, or where mitigation may not work, this is made clear and data are introduced to justify the acceptance of these assumptions.				
	4.2	3.2.1b, 2.5.1	Indicates the significance of any residual or unmitigated impacts remaining after mitigation, and justifies why these impacts should not be mitigated.				
Commitment to mitigation and monitoring							
	4.7	3.3.1	Gives details of how the mitigation measures will be implemented and function over the time span for which they are necessary.				
	4.8	3.3.2	Proposes monitoring arrangements for all significant impacts, especially where uncertainty exists, to check the environmental impact resulting from the implementation of the project and their conformity with the predictions made.				
	4.9		The scale of any proposed monitoring arrangements corresponds to the potential scale and significance of deviations from expected impacts.				
Environmental effects of mitigation							
	4.10		Investigates and describes any adverse environmental effects of mitigation measures.				
	4.11		Considers the potential for conflict between the benefits of mitigation measures and their adverse impacts.				

Overall mark:

7. NON-TECHNICAL SUMMARY

min	EU	LC	criterion	criteria			comments
	IAU			√ x	IAU/LC	EU	
	Non-technical summary						
*	6.1 4.	4.1a, 4.4.2b	There is a non-technical summary of the main findings of the study, which contains at least a brief description of the project and the environment, an account of the main mitigation measures to be undertaken by the developer, and a description of any remaining or residual impacts.				
	6.2	4.4.1b	The summary avoids technical terms, lists of data and detailed explanations of scientific reasoning.				
*	6.3	4.4.2a	The summary presents the main findings of the assessment and covers all the main issues raised in the information.				
	6.4	4.4.2c	The summary includes a brief explanation of the overall appoach to the assessment.				
	6.5	4.4.2d	The summary indicates the confidence which can be placed in the results.				

Overall mark:

8. ORGANISATION AND PRESENTATION OF INFORMATION

min	EU	LC	criterion	performance against criteria			comments
	IAU			√ x	IAU/LC	EU	
Organisation of the information							
	8.1	4.1.2a	Logically arranges the infomation in sections.				
	8.2	4.1.2b	Identifies the location of information in a table or list of contents.				
		4.1.3	There are chapter or section summaries outlining the main findings of each phase of the investigation.				
	8.3	4.1.4	When information from external sources has been introduced, a full reference to the source is included.				
Presentation of information							
			Mentions the relevant EIA legislation, name of the developer, name of competent authority(ies), name of organisation preparing the EIS, and name, address and contact number of a contact person.				
		4.1.1	Includes an introduction briefly describing the project, the aims of the assessment, and the methods used.				
		4.2.3	The statement is presented as an integrated whole. Data presented in appendices are fully discussed in the main body of the text.				
	8.4		Offers information and analysis to support all conclusions drawn.				
	8.5	4.2.1	Presents information so as to be comprehensible to the non specialist. Uses maps, tables, graphical material and other devices as appropriate. Avoids unnecessarily technical or obscure language.				
	8.6		Discusses all the important data and results in an integrated fashion.				
	8.7	4.3.1b	Avoids superfluous information (i.e. information not needed for the decision).				
	8.8		Presents the information in a concise form with a consistent terminology and logical links between different sections.				
	8.9	4.3.1a	Gives prominence and emphasis to severe adverse impacts, substantial environmental benefits, and controversial issues.				
		4.2.2	Defines technical terms, acronyms and initials.				
	8.10	4.3.2	The information is objective, and does not lobby for any particular point of view. Adverse impacts are not disguised by euphemisms or platitudes.				
Difficulties compiling the information							
	7.1		Indicates any gaps in the required data and explains the means used to deal with them in the assessment.				
	7.2		Acknowledges and explains any difficulties in assembling or analysing the data needed to predict impacts, and any basis for questioning assumptions, data or information.				

Overall mark:
COLLATION

Simple "regulatory requirements" criteria (from min. column; x/√):
describes development (EU 1.4) ___
defines land area used (1.8, 1.9) ___
describes land uses (1.10) ___
considers direct... effects (5.1, 5.2) ___
impacts on human beings etc. (5.3a) ___
considers mitigation (4.1a) ___
mitigation includes... (4.4) ___
there is a non-technical summary (6.1) ___
the summary covers ... (6.3) ___

covers criteria (yes/no): ___

IAU criteria:
1. Description of the development ___
2. Description of the environment ___
3. Scoping, consultation, and impact identification ___
4. Prediction and evaluation of impacts ___
5. Alternatives ___
6. Mitigation and monitoring ___
7. Non-technical summary ___
8. Organisation and presentation of information ___

Overall mark (A-F): ___

Overall quality (good and bad points) of the EIS for:

developer:

LA:

consultant:

public:

statutory consultees:

other:

APPENDIX 6: LEE AND COLLEY CRITERIA AND EU CRITERIA

A6.1 Lee and Colley criteria

Overall the results of this section are very similar to those for the IAU criteria detailed in Chapter 4.

Pre-1991: Table A6.1 summarises how the pre-1991 ESs covered the Lee and Colley criteria. The overall quality was just unsatisfactory (D), with 8 (32%) poor ESs (D/E to F), 13 (52%) marginal ESs (C to D), and 4 (16%) good ESs (A to B/C). Alternatively, 10 (40%) of the ESs were satisfactory (A to C), 14 (56%) were unsatisfactory (D to F), and 1 (4%) was in between (C/D). Within the ES, the description of the development and the environment, and the communication of results were carried out slightly better (C/D), than impact identification and assessment, and alternatives and mitigation (D).

Post-1991: Table A6.2 summarises how the post-1991 ESs covered the Lee and Colley criteria. The overall quality was just satisfactory (C), with 4 (16%) poor ESs, 11 (44%) marginal ESs, and 10 (40%) good ESs. Alternatively, 14 (56%) of the ESs were satisfactory, 7 (28%) were unsatisfactory, and 4 (16%) were in between. The description of the development and the environment, impact identification and assessment, and the communication of results were generally carried out satisfactorily (C), whilst alternatives and mitigation were carried out less well (C/D).

Changes over time: Figure A6.1 confirms that there has been a significant improvement in the quality of ESs between 1988-90 and 1992-94, from an average of just unsatisfactory (D) to just satisfactory (C). According to the Lee and Colley criteria, the proportion of poor ESs (D/E to F) halved from 32% to 16%, and the proportion of good ESs (A to B/C) more than doubled from 16% to 40%. The proportion of satisfactory (A to C) ESs rose from 40% to 56%, and the proportion of unsatisfactory (D to F) ESs halved, from 56% to 28%. The range of quality remained unchanged, ranging from B to F.

Quality improved in each of the 4 main categories of assessment, with the description of the development and environment, and ES presentation improving from C/D to C, alternatives and mitigation improving from D to C/D, and identification and evaluation of key impacts improving from D to C.

Figure A6.2 shows the amount of change over time within each pair. Sixteen pairs showed an improvement over time, and 9 showed the reverse. Fourteen of the ESs improved by between 0.5 and 2 marks, one pair of waste disposal ESs improved by 3 marks, and the motorway service area ESs improved by 4 marks. Seven became worse by 0.5 or 1 mark, one by 1.5 marks, and one pair of waste disposal ESs worsened by 2 marks.

Table A6.1 Lee and Colley criteria: pre-1991 ESs (25 ESs)

Criterion / Quality	A	A/B	B	B/C	C	C/D	D	D/E	E	E/F	F	total
1. Description of the development, local environment, & baseline conditions		1	4	1	7	1	5		3	1	2	C/D
2. Identification & evaluation of key impacts			2	1	7	1	5	1	4		4	D
3. Alternatives & mitigation		2	2	4		4	3	8		2		D
4. Communication of results		2	3	1	3	3	8	2	2	1		C/D
Overall ES*			4		6	1	6	3	2	1	2	D
			good = 4			marginal = 13			poor = 8			
			satisfactory = 10				1		unsatisfactory = 14			

*Average of 4 criteria

Table A6.2 Lee and Colley criteria: post-1991 ESs (25 ESs)

Criterion / Quality	A	A/B	B	B/C	C	C/D	D	D/E	E	E/F	F	total
1. Description of the development, local environment, & baseline conditions			8	2	8	2	3		1		1	C
2. Identification & evaluation of key impacts			6		10	1	4	1	2		1	C
3. Alternatives & mitigation			5	1	6	2	5	1	4	1		C/D
4. Communication of results		1	6	2	6	2	5		2	1		C
Overall ES*			5	5	4	4	3	2	1		1	C
			good = 10			marginal = 11			poor = 4			
			satisfactory = 14				4		unsatisfactory = 7			

*Average of 4 criteria

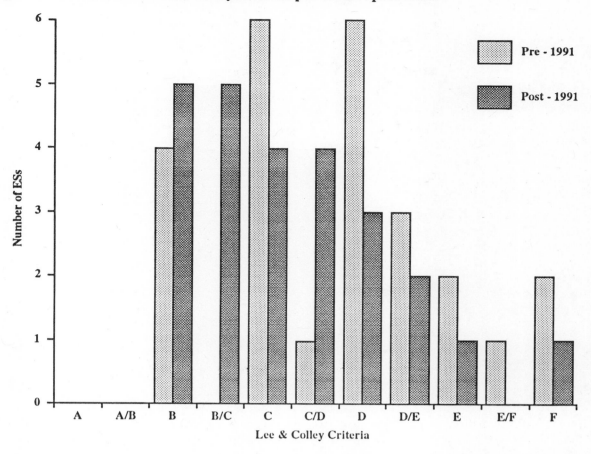

Figure A6.1: Marks for Lee & Colley criteria: pre-1991 v. post-1991.

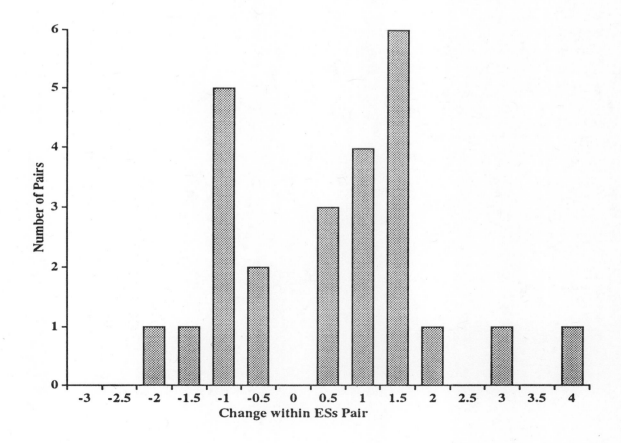

Figure A6.2: Change in ES quality (Lee and Colley) within pairs: pre-1991 v. post-1991

A6.2 EU criteria

Pre-1991: Table A6.3 summarises how the pre-1991 ESs covered the EU criteria. Overall, almost half (48%) of the ESs were incomplete, with 9 (36%) being acceptable and 4 (16%) complete. Generally the individual components of the ESs followed this trend, with alternatives being considered least well.

Post-1991: Table A6.4 summarises how the post-1991 ESs covered the EU criteria. Overall, 5 ESs were complete, 11 were acceptable, and 9 were incomplete. Again, alternatives were considered least well.

Changes over time: Figure A6.3 shows that there seems to have been a significant improvement in the quality of ESs between 1988-90 and 1992-94, from an average of between acceptable and incomplete to an average of just less than acceptable. The number of incomplete ESs fell from 12 to 8, and the number of complete ESs rose from 4 to 5.

Quality improved in each of the 8 categories of assessment, with a significant drop in "incomplete" marks. The prediction and evaluation of impacts, and the consideration of alternatives in particular, improved. Figure A6.4 shows the amount of change over time within each pair. Using the reviewers' judgement as a criterion, 11 improved over time and 5 got worse. Using the median as the criterion, 12 improved and 5 got worse.

Table A6.3: EU criteria: pre-1991 ESs (25 ESs)

Criterion	Comp.	Accept.	Incomp.
1. Description of the devel.	5	11	9
2. Description of the env.	5	11	9
3. Scoping, consultation, & impact identification	2	12	11
4. Prediction & evaluation of impacts	3	9	13
5. Alternatives	2	3	20
6. Mitigation & monitoring	6	7	12
7. Non-technical summary	6	6	13
8. Organisation & presentation of info.	6	9	10
Overall ES: . reviewer's judgement	3	9	13
. median	4	10	11

Table A6.4: EU criteria: post-1991 ESs (25 ESs)

Criterion	Comp.	Accept.	Incomp.
1. Description of the devel.	5	11	9
2. Description of the env.	5	11	9
3. Scoping, consultation, & impact identification	2	12	11
4. Prediction & evaluation of impacts	3	9	13
5. Alternatives	2	3	20
6. Mitigation & monitoring	6	7	12
7. Non-technical summary	6	6	13
8. Organisation & presentation of info.	6	9	10
Overall ES: . reviewer's judgement	3	9	13
. median	4	10	11

Figure A6.3: Grades for EU criteria: pre 1991 v. post-1991.

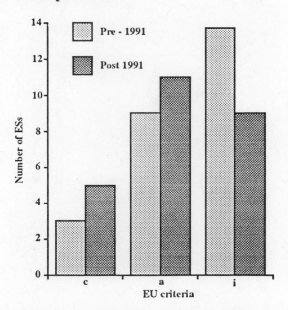

Figure A6.4: Changes in ES quality (EU criteria) within pairs: pre-1991 v. post-1991.

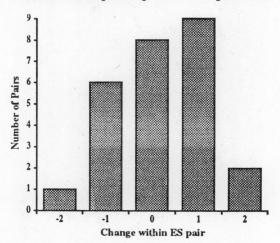

APPENDIX 7: 'QUALITY FOR WHOM?' CASE STUDIES: CHECKLIST AND QUESTIONNAIRE

Research team should complete as much of Section 1(a)-1(e) as possible from the case background papers and the EIS. Section 2 provides the broad template for individual party interviews (by telephone and/or direct) as needed.

SECTION 1: CASE STUDY CHECKLIST

1(a) Details of project

1	Project number	
2	Project type	
3	Project size (ha,km)	
4	Application status (outline/full)	
5	Regs. Schedule number	
6	Determining authority (County/District)	

1(b) Details of application process

7	Date of application	
8	Applicant	
9	EIS prepared by	
10	Need for EIS requested/volunteered	
11	Was need for EIS subject to appeal (include dates)?	
12	Was EIS submitted after the planning application?	
13	List consultees and bodies which responded to consultation process	
14	Were other (non-planning) departments within the authority consulted (list which)	
15	List main issues identified by EIS	
16	List any issues not identified by EIS but considered significant by LPA or consultees	
17	Did the LPA request further information (specify information requested, relevant dates and whether further information supplied)?	
18	How was the EIS presented to Council Members (officer's summary, EIS available etc)?	
19	LPA decision and date, and where relevant date and decision of appeal or 'call in' procedure.	
20	Has the project commenced or been completed?	

1(c) Quality for whom issues

21	Agreed Lee and Colley grade for 4.3.2	
22	Did EIS provide minimum regulatory requirements?	
23	Size of EIS (number of volumes, number of pages for main text and for non-technical summary.	
24	List bodies, including the LPA consulted prior to the production of the EIS.	
25	Was the EIS the subject of any pre-application public participation exercise?	
26	Number of EISs previously received by the LPA.	
27	Number of EISs previously submitted by the developer.	
28	Number of EISs previously prepared by the consultant.	
29	Was the EIS subject to independent review (by whom and which review system used)?	
30	Was the EIS prepared on the basis of any published guidance (state which)?	
31	Were the contents and findings of the EIS questioned or criticised by any party during the application process (state which body and what the issues were)?	

1(d) Key Participants in the application process.

From the background papers list the names of individuals and organisations involved in the process.

1(e) List all unresolved questions from 1 to 31 above and identify the individual or organisation from which additional information is needed.

SECTION 2: INTERVIEWEE QUESTIONNAIRE

2(a) Context

1.	Project reference number	
2.	Name of interviewee	
3.	Organisation	
4.	What was your role in the processing of the EIS and planning application?	
5.	When did you first become involved in the process?	
6.	Was this the first time you had been involved with a project which needed an EIS?	

2(b) Views on quality

7.	In your view did the EIS fulfil the minimum requirements of the regulations?			
8.	What is your overall assessment of the quality of the EIS?	Good	Marginal	Poor
9.	On what basis do you make this assessment?			
10.	Did you use any structured review system or guidance to judge the quality of the EIS? If yes, which?			
11.	To what extent did the EIS assist in making the decision on the planning application?	Much	Some	Little/None
12.	How would you assess the EIS against the following criteria? • Comprehensiveness • Objectivity • Clarity of information	Good	Marginal	Poor
13.	(If possible) how well does the EIS cover the following? • Description of development • Description of environment • Scoping, consultation and impact identification • Prediction and evaluation of impacts • Alternatives • Mitigation and monitoring • Non-technical summary • Organisation and presentation of information	Good	Marginal	Poor

2(c) Costs

14.	. How much time did your authority/firm etc spend on the EIS? . How was this costed? . Costs by grade of staff? . Other running costs?	
15..	. How much time did others employed by your authority/firm spend on the EIS? . How was this costed? . Costs by grade of staff? . Other running costs?	
16.	. What other costs were associated with your work on the EIS? e.g. . Any capital purchases; were these solely for the EIS, or do they have a residual value? . Other work delayed? . Others?	
17.	. What were the benefits of the EIS process; and can these be quantified? e.g. . Time savings . Design savings . PR value? . Improvement in quality of project? . Protection of particular environmental resources? . Others?	